# Stem Cell Biology and Regenerative Medicine

**Series Editor**

Kursad Turksen, Ph.D.
kursadturksen@gmail.com

Our understanding of stem cells has grown rapidly over the last decade. While the apparently tremendous therapeutic potential of stem cells has not yet been realized, their routine use in regeneration and restoration of tissue and organ function is greatly anticipated. To this end, many investigators continue to push the boundaries in areas such as reprogramming, stem cell niche, nanotechnology, biomimetics and 3D bioprinting, to name just a few. The objective of the volumes in the *Stem Cell Biology and Regenerative Medicine* series is to capture and consolidate these developments in a timely way. Each volume is thought-provoking in identifying problems, offering solutions, and providing ideas to excite further innovation in the stem cell and regenerative medicine fields.

More information about this series at http://www.springer.com/series/7896

Kursad Turksen
Editor

# Autophagy in Health and Disease

## Potential Therapeutic Approaches

Humana Press

*Editor*
Kursad Turksen
Ottawa, ON
Canada

ISSN 2196-8985 ISSN 2196-8993 (electronic)
Stem Cell Biology and Regenerative Medicine
ISBN 978-3-030-07457-9 ISBN 978-3-319-98146-8 (eBook)
https://doi.org/10.1007/978-3-319-98146-8



This Humana Press imprint is published by the registered company Springer Nature Switzerland AG
The registered company address is: Gewerbestrasse 11, 6330 Cham, Switzerland

# Preface

The importance of autophagy in the maintenance and self-renewal of stem cell populations and, more generally, tissue homeostasis is well recognized. However, understanding how autophagy achieves its effects on stem cells and their progeny in both normal and disease states is less advanced and thus an area of very active investigation. The objective of this volume is to provide an up-to-date overview of the current state of knowledge in the area and where the field may go from here.

I would like to acknowledge that the idea for this volume resulted from a discussion that I had with Christina Dzikowski, Publisher at Springer Nature, and I remain grateful for her encouragement. I am also grateful to Patrick Marton, Executive Editor of Springer Protocols, for introducing me to Christina.

I would also like to acknowledge and thank Seth Fried at Springer Nature for his outstanding help and support throughout this project and the series to which it belongs.

A special thank you goes to Bibhuti Sharma and the production crew for all of their work on the volume.

Finally, I thank all of the contributors for their generosity in giving their expertise and time to the chapters of this volume. Without them, the volume would not have materialized as an important synthesis of where the field of autophagy and stem cells is and is going.

Ottawa, ON, Canada　　Kursad Turksen

# Contents

# Contributors

**Toshihiko Aki** Department of Forensic Medicine, Tokyo Medical and Dental University, Tokyo, Japan

**Marta E. Alvarez-Argüelles** Servicio de Microbiología, Hospital Universitario Central de Asturias, Oviedo, Spain

Instituto de Investigación Sanitaria del Principado de Asturias, ISPA, Oviedo, Spain

**Nebojsa Arsenijevic** Faculty of Medical Sciences, Department of Microbiology and Immunology, Center for Molecular Medicine and Stem Cell Research, University of Kragujevac, Serbia, Kragujevac, Serbia

**Manmohan Bajaj** Stem Cell Lab, National Centre for Cell Science, Pune, India

**Shalmoli Bhattacharyya** Department of Biophysics, PGIMER, Chandigarh, India

**Cigir Biray Avci** Department of Medical Biology, Faculty of Medicine, Ege University, Izmir, Turkey

**José A. Boga** Servicio de Microbiología, Hospital Universitario Central de Asturias, Oviedo, Spain

Instituto de Investigación Sanitaria del Principado de Asturias, ISPA, Oviedo, Spain

**Zhiyong Cheng** Department of Human Nutrition, Foods and Exercise, Fralin Life Science Institute, College of Agriculture and Life Science, Virginia Tech, Blacksburg, VA, USA

**Ana Coto-Montes** Departamento de Morfología y Biología Celular, Universidad de Oviedo, Oviedo, Spain

Instituto de Investigación Sanitaria del Principado de Asturias, ISPA, Oviedo, Spain

**Crissy Fellabaum** Regenerative Processing Plant, LLC, Palm Harbor, FL, USA

**Bakiye Goker Bagca** Department of Medical Biology, Faculty of Medicine, Ege University, Izmir, Turkey

**C. Randall Harrell** Regenerative Processing Plant, LLC, Palm Harbor, FL, USA

**Daisuke Hirayama** Department of Gastroenterology and Hepatology, Sapporo Medical University School of Medicine, Sapporo, Hokkaido, Japan

**Tomoya Iida** Department of Gastroenterology and Hepatology, Sapporo Medical University School of Medicine, Sapporo, Hokkaido, Japan

**Jelena Jakovljevic** Faculty of Medical Sciences, Department of Microbiology and Immunology, Center for Molecular Medicine and Stem Cell Research, University of Kragujevac, Serbia, Kragujevac, Serbia

**Vaijayanti P. Kale** Stem Cell Lab, National Centre for Cell Science, Pune, India

Symbiosis Centre for Stem Cell Research, Symbiosis School of Biological Sciences, Symbiosis International (Deemed University), Pune, India

**Rohan S. Kulkarni** Stem Cell Lab, National Centre for Cell Science, Pune, India

**Ajay Kumar** Department of Biophysics, PGIMER, Chandigarh, India

**Kai Li** College of Medicine, State Key Laboratory of Medicinal Chemical Biology, Nankai University, Tianjin, China

**YongTian Liang** Institute of Biology/Genetics, Free University of Berlin, Berlin, Germany

NeuroCure, Cluster of Excellence, Charity Universitätmedizin, Berlin, Germany

**Longhua Liu** Department of Pathology and Cell Biology, Columbia Diabetes Research Center, New York Obesity Research Center, Columbia University, New York, NY, USA

Department of Human Nutrition, Foods, and Exercise, Fralin Life Science Institute, College of Agriculture and Life Science, Virginia Tech, Blacksburg, VA, USA

**Hiroshi Nakase** Department of Gastroenterology and Hepatology, Sapporo Medical University School of Medicine, Sapporo, Hokkaido, Japan

**Kanako Noritake** Department of Forensic Medicine, Tokyo Medical and Dental University, Tokyo, Japan

**Yasunao Numata** Department of Gastroenterology and Hepatology, Sapporo Medical University School of Medicine, Sapporo, Hokkaido, Japan

**Zulema Pérez-Martínez** Servicio de Microbiología, Hospital Universitario Central de Asturias, Oviedo, Spain

Instituto de Investigación Sanitaria del Principado de Asturias, ISPA, Oviedo, Spain

**Russel J. Reiter** Department of Cellular and Structural Biology, UT Health Science Center at San Antonio, San Antonio, TX, USA

**Koichi Uemura** Department of Forensic Medicine, Tokyo Medical and Dental University, Tokyo, Japan

**Kana Unuma** Department of Forensic Medicine, Tokyo Medical and Dental University, Tokyo, Japan

**Vladislav Volarevic** Faculty of Medical Sciences, Department of Microbiology and Immunology, Center for Molecular Medicine and Stem Cell Research, University of Kragujevac, Serbia, Kragujevac, Serbia

**Irina Voronov** Faculty of Dentistry, University of Toronto, Toronto, ON, Canada

**Kohei Wagatsuma** Department of Gastroenterology and Hepatology, Sapporo Medical University School of Medicine, Sapporo, Hokkaido, Japan

**Zhuo Yang** College of Medicine, State Key Laboratory of Medicinal Chemical Biology, Nankai University, Tianjin, China

**Ralph A. Zirngibl** Faculty of Dentistry, University of Toronto, Toronto, ON, Canada

# Chapter 1
# Autophagy and Stem Cells

**Kai Li and Zhuo Yang**

**Abstract** Autophagy, as a highly conserved cellular process, can achieve the degradation and recycling of intracellular substances, and is crucial for maintaining cellular homeostasis and remodeling of normal development. Dysfunctions in autophagy would cause a variety of illnesses including cancer, inflammatory bowel disease and neurodegenerative diseases. The unique self-renewal ability and differentiation ability of stem cells can improve these diseases. Therefore, exploring the mechanism of autophagy in maintaining stem cell homeostasis is crucial. Here we review the mechanisms and regulation of autophagy in embryonic stem cells, hematopoietic stem cells, mesenchymal stem cells, neural stem cells, and cancer stem cells. It helps us understand the relationship between autophagy and stem cells. Although there are many unanswered questions, the study of autophagy and stem cell biology can help us to progress in life sciences.

## Abbreviations

| | |
|---|---|
| ASCs | Adult stem cells |
| ATG | Autophagy-related |
| BM-MSCs | Bone marrow MSCs |
| CMA | Chaperone-mediated autophagy |
| ESCs | Embryonic stem cells |
| HSCs | Hematopoietic stem cells |
| HSPCs | HSCs and progenitor cells |
| iPS | Induced pluripotent stem |
| LAMP2 | Lysosomal-associated membrane protein 2 |
| MSCs | Mesenchymal stem cells |
| NSCs | Neural stem cells |

K. Li · Z. Yang (✉)
College of Medicine, State Key Laboratory of Medicinal Chemical Biology, Nankai University, Tianjin, China
e-mail: Zhuoyang@nankai.edu.cn



K. Turksen (ed.), *Autophagy in Health and Disease*, Stem Cell Biology and Regenerative Medicine, https://doi.org/10.1007/978-3-319-98146-8_1

| | |
|---|---|
| ROS | Reactive oxygen stress |
| SD-MSCs | Serum-deprived MSCs |
| SGZ | Subgranular zone |
| SVZ | Subventricular zone |

## 1.1 Types and Functions of Autophagy

The concept of autophagy is first proposed by the Belgian cytology and biochemist Christian de Duve at the International Symposium on Lysosomes in Paris [1]. Autophagy comes from the Greek words "auto" and "phagy," meaning "self" and "swallowed," respectively. Autophagy therefore is a "self-eating" process in cells that distinguishes itself from "self-killing" phenomenon of apoptosis. Over the years, it has been found that autophagy is a highly conserved cellular process of renewal, clearance and repair in cells, and its main physiological function is transport of long-lived proteins, damaged and dysfunctional organelles to lysosomes for degradation and mobilization of intracellular nutrients in order to maintain the basal energy balance [2, 3]. At this stage, research has shown that autophagy is essentially an effective feedback of cells in the face of environmental changes and plays an important role in the metabolism. Firstly, autophagy can ensure cell survival under the stress conditions such as starvation, energy loss, and reactive oxygen stress (ROS) in eukaryotes. Under starvation conditions, cells degrade proteins, carbohydrates, and lipids by inducing autophagy, thereby releasing nucleotides, amino acids, and free fatty acids to maintain the energy status of cells [4]. Secondly, in eukaryotic cells, autophagy, as a natural defense mechanism, can not only degrade the intracellular damaged components and promote cell survival, but is also a cell death mechanism, which is parallel with apoptosis and necrosis inducing cell death [3].

### *1.1.1 Classification and Function of Autophagy*

Three kinds of autophagy have been characterized to date. (1) Macroautophagy refers to degradation of long-lived proteins and necrosis of organelles wrapped by specialized double-membraned vesicles known as autophagosomes which are then carried to lysosomes. (2) Microautophagy depends on lysosomal membrane invagination, prominent, separated, and then packed cytoplasmic compounds. (3) Chaperone-mediated autophagy (CMA) is a complex pathway in which proteins are directly targeted from the cytosol to the lysosomes [5]. CMA involves the lysosomal-associated membrane protein 2 (LAMP2)-dependent translocation of autophagic substrates bound to cytosolic chaperones of the heat shock protein family across the lysosomal membrane [6]. Autophagy is commonly referred to as macroautophagy, and this review focuses on the role of macroautophagy (referred to as autophagy) in stem cells.

### 1.1.2 Characteristics of Autophagy

The process of autophagy is a complete process involving a series of autophagic structures and mainly consists of five successive subtle steps (Fig. 1.1) including the activity of related regulatory components in mammalian cells [7]. The five steps include initiation of autophagy, vesicle nucleation, vesicle elongation and completion, fusion and degradation, and termination of autophagy [8, 9].

In the initiation step of autophagy, a variety of factors, such as nutrient or energy starvation, ROS accumulation and exposure to rapamycin, can induce the initiation of autophagy by activating serine-threonine protein kinase ULK1 [2]. Then the ULK1 complex is formatted by controlling the sequential binding of a series of proteins including ULK1, ATG13, ATG101, and FIP200 and then it is dissociated from the mTORC1 complex [10–12] for commencing the next step of autophagy.

In second step, vesicle nucleation and the complex protein of Beclin-1-Vps34-Vps15-ATG14L is involved [13]. The activated ULK1 is bound to the complex protein Beclin-1-Vps34 through phosphorylation of Ambra1 [14]. Thereafter Vps34 interacts with an effector protein DFCP1 and the protein of WIPI to generate the phosphatidylinositol-3-phosphate (PI3P) to initially form autophagosomes [15, 16].

In the vesicle elongation, the autophagy-related (ATG) proteins participate in the expansion and completion of the autophagosome. In this section, on the one hand, ATG12 sequentially interacts with ATG7/ATG10 to form the ATG12-ATG5 protein

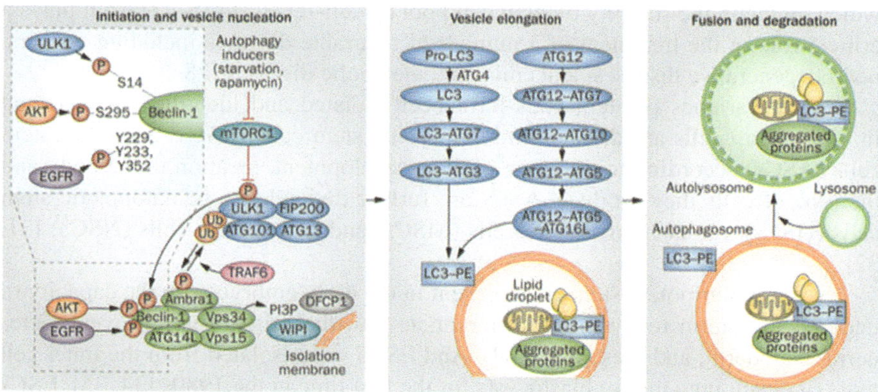


**Fig. 1.1** The process of autophagy involves multiple proteins and signaling pathways. In the initial process, autophagic stimuli such as nutrient deprivation or rapamycin activate the protein complex of ULK1-ATG101-ATG13-FIP200 and thereby format the complex of Beclin-1-Vps34 by phosphorylating Ambra1 that regulate the activity of ULK1 through interaction with TRAF6 and Beclin-1 that is independent of mTORC1, in which are responsible for promoting autophagosome. The vesicle elongation of autophagosome is mediated by two conjugated systems consisting of ATG12-ATG5-ATG16L and LC3-PE. With the formation of complete autophagic vesicles, the mature autophagosome fuses with a lysosome to produce an autolysosome that can degrade cytoplasmic compounds or organelles. Permission obtained from Springer Nature © Kim, K.H. and M.S. Lee-Nature Reviews Endocrinology, 2014. 10(6): p. 322

complex that then conjugates to ATG16L, forming the ATG12-ATG5-ATG16L complex [17–19]. On the other hand, LC3 produced by Pro-LC3 recruits a protease of ATG4. Then LC3 is bound to ATG7/ATG3, forming the LC3-ATG3 complex [20, 21]. Thereupon the LC3-ATG3 complex conjugates phosphatidylethanolamine and the ATG12-ATG5-ATG16L complex to extend and close the autophagosome double membranes [22].

The next step is fusion and degradation, in which target proteins are tethered or organelles are encapsulated by mature autophagosomes that are fused with lysosomes to form autophagolysosomes [23].

At the termination of autophagy, the production of nutrients and energy by the autolysosomes reactivates mTOR to attenuate autophagy, and which is followed by a feedback mechanism that inhibits the excessive activation of autophagy during periods of starvation or ROS [24].

In summary, the process of autophagy is completed under the complex and orderly fine regulation of a variety of proteins.

## 1.2 Types and Characteristics of Stem Cells

Stem cell biology has attracted tremendous interest recently. Stem cells are a group of non-terminally differentiated cells with self-renewal, multidirectional differentiation potential, and quiescence to produce more stem cells or specialized tissues, which maintain the stability of stem cell pool by self-renewal [25–27], so it plays a critical role in the treatment of a number of incurable diseases including cancers, neurodegenerative diseases, leukemia, and metabolic diseases [28–30].

Several varieties of stem cells have been isolated and identified in vivo and in vitro. Stem cells are divided into embryonic stem cells (ESCs) and adult stem cells (ASCs) according to the stage of their development, location in the body and the type of cells they produce. ASCs are further divided into hematopoietic stem cells (HSCs), mesenchymal stem cells (MSCs) and neural stem cells (NSCs) [31, 32].

ESCs are pluripotent stem cells present in the early embryo with the capacity to undergo long-term renewal and differentiate into the primary germ layers: ectoderm, endoderm, and mesoderm [33]. And ESCs were isolated from the inner cell mass of developing mouse blastocysts for the first time in the 1980s [34, 35]. ESCs have the ability to proliferate without differentiation in vitro. After being injected into the body to form a chimera with the intact embryo, ESCs can develop a series of adult tissues including germ cells [36]. When under suitable culture conditions, ESCs can differentiate to various cell types. Therefore, exploring how autophagy precisely regulates the proliferation and differentiation of ESCs is the primary point of current researches.

HSCs are multipotent cells that have the highly potential to self-renew, self-sustaining (that is, producing hematopoietic progenitor cells while keeping their own number constant) and differentiate into all mature blood cell types [33, 37, 38], and are derived early in embryogenesis from mesoderm and reside in the bone marrow throughout the lifetime of the animal [39, 40]. In addition, HSCs have attracted more and more researchers in recent years because it is easily separated from bone marrow and blood.

MSCs are a population of multipotent cells that have self-replication ability and multidirectional differentiation potential, and are capable of differentiating into a variety of cell lineages such as bone, muscle, osteocytes, cartilage, adipocytes, chondrocytes, and stromal cells under appropriate conditions in vitro and in vivo [41–45]. And it has demonstrated that MSCs have an effect on hematopoiesis, immune response, and treatment of neurodegenerative diseases [45, 46].

NSCs have the potential to differentiate into a variety of cells. Studies have found that NSCs can be differentiated into different types of neurons, astrocytes, oligodendrocytes, and other types of cells under different conditions [47, 48]. And the same source of NSCs can be differentiated into different nerve cells and even be differentiated into myocytes, epithelial cells, and hematopoietic cells in different environments [49, 50]. Thus NSCs mainly include the following features: (1) can proliferate and differentiate into neural tissue or are derived from the nervous system; (2) have the ability of self-renewal ability; (3) can promote other cells through asymmetric cell division [51].

In this review, we describe a growing body of knowledge encompassing a range of stem cell systems that have significantly advanced our understanding of autophagy in stem cell biology.

## 1.3 The Regulation of Autophagy in Different Stem Cells

Autophagy, as a crucial mechanism of maintaining cellular homeostasis in cells, not only plays a pivotal in "remodeling" but also eliminates dysfunctional cells, so that it has been associated with different physiological and pathological conditions such as cancer, neurodegenerative disease, and aging et al. [24, 33, 52]. Moreover, as stem cells have the characteristics of pluripotency, proliferation, and self-renewal, it is crucial to precisely regulate and maintain the homeostasis of stem cells. Recent publications have shown that these characteristics (self-renewal, pluripotency, differentiation, and quiescence) of stem cells are dependent on the process of autophagy [52, 53]. Thus autophagy has been implicated that regulates the development and differentiation of a variety of stem cells including clearance of mitochondria and other organelles during erythrocyte maturation reported decades ago [54, 55] and recently validated by gene knockout studies [56].

### 1.3.1 The Role of Autophagy in Embryonic Stem Cells

ESCs, as pluripotent stem cells, are derived from totipotent cells of the early mammalian embryo and have the ability to proliferate and differentiate in vitro [36, 57]. More and more research works have focused on ESCs and have found that autophagy affects the development of ESCs, which would be very helpful for future research [58]. The initial study of autophagy was mainly founded on yeast *Saccharomyces cerevisiae*, in which the ATG genes was identified [59]. As the awareness of autophagy deepened, it crossed from yeast to mammals. For example, the first study of the molecular mechanism of autophagy in mammalian cells was performed by mouse embryonic stem cells [60]. Although studies on embryonic stem cells span from yeast to mammals, many molecular mechanisms remain unclear. Noteworthily, existing studies have shown that autophagy has an impact on early embryonic development and late development. On the one hand, it has also been shown that autophagy plays a crucial role in early stage of embryonic development [61, 62]. Since the development and differentiation of embryos are usually accompanied by remodeling of cells and tissues, which implies that autophagy plays a considerable role in this process, providing the raw material for building new structures and removing the old materials [33, 63]. For example, it was reported that basal level of autophagy was detected in undifferentiated human ESCs, but the pluripotent proteins of Oct4, Sox2 and Nanog were found to accumulate in human ESCs by incubating with the inhibition of autophagy such as BafA1 and 3-MA. This suggest that pluripotent protein could be degraded by autophagy [64]. In addition, Mizushima N et al. have found that the intracellular bulk protein turnover is significantly reduced when mouse ESC *atg5*$^{-/-}$ cells are cultured in medium, in which amino acids have been labeled with [$^{14}$C] [60]. During the differentiation of human ESCs, the autophagy had been significantly activated when human ESCs were cultured in medium with TGF-β receptor inhibitor or removing the maintenance factor secreted by fibroblasts [65]. Besides, after removal of leukemia inhibitory factor, ESCs did not form normal embryoid bodies in *atg5*$^{-/-}$ mouse [66]. The above studies show that autophagy takes part in the regulation of protein homeostasis and plays an important role in maintaining the pluripotency of ESCs in the early stages of embryogenesis.

On the other hand, although autophagy has an impact on the early stages of embryogenesis, it has little effect on the later stages of embryogenesis. For example, Mizushima and Levine et al. have reported that *atg3*$^{-/-}$, *atg5*$^{-/-}$, *atg7*$^{-/-}$, *atg9*$^{-/-}$, or *atg16l1*$^{-/-}$ mice are not embryonically lethal [4, 66–69], although they die within 1–2 days of birth accompanying with weight loss, in which presumably could due to suckling defects caused by deficient neurological development [70]. In addition, Mizushima and Levine et al. and Tsukamoto et al. have thought that the reason for neonatal survival of these embryos due to maternally inherit the ATG protein that come from the oocyte cytoplasm [61, 70, 71]. However, in contrast to other *atg*$^{-/-}$

mouse models, mice that are null for encoding the phagophore-forming proteins of BECN1, AMBRA1, or FIP200 are embryonically lethal [72–74]. The mechanism of this is unclear. But researchers speculate that the other functions of such proteins may be affected due to knocking out this type of gene. For example, Chen et al. have demonstrated that the FIP200 gene has nonautophagic function with sustaining embryogenesis [73].

In addition, autophagy has been shown to promote morphological changes in stem cells [53]. For example, Kuo et al. have shown that starvation or treatment with rapamycin activate the process of autophagy and promote the degradation of the midbody during embryonic stem cell differentiation [75]. Midbodies are a circular structure that forms cytoplasmic bridges after cytokinesis and are necessary for the isolation of daughter cells [76]. Isakson et al. have discovered that midbody can be eliminated by midbophagy involving a complex consisting of p62 (SQSTM1), ALFY (WDFY3), and TRAF6 [77], and interactions between NBR1 and CEP55 [75] and participation of TRIM17 [78]. And Kuo et al. have pointed out that the elimination of midbodies undergoes differentiation [75]. And it has been suggested that correct control of midbophagy would be essential to maintain the pluripotency of ESCs.

In summary, autophagy affects the differentiation and proliferation of ESCs throughout the embryonic stage.

## 1.3.2 The Effect of Autophagy on Hematopoietic Stem Cells

### 1.3.2.1 The Role of Autophagy in Hematopoietic Stem Cells

Since HSCs are easily obtained from bone marrow and blood, the research on HSCs is relatively abundant at this stage. Many studies have investigated that autophagy plays an important regulatory role in the self-renewal, proliferation, and differentiation of HSCs. Furthermore, some studies have suggested that the inhibition of autophagy can cause myeloproliferative disorders in HSCs. In *atg7*$^{-/-}$ HSCs, the abnormally increased number of mitochondria and the impaired function of mitophagy, resulting in a large number of ROS and damaging-DNA accumulated in cells [79]. In addition, the remove of cytokines and energy made *atg12*$^{-/-}$ HSCs prone to apoptosis, whereas wild-type HSCs are protected by FoxO3A-mediated autophagy [80]. Then HSCs derived from aged mice have shown that regenerative activity is decreased and self-renewal is diminished in increasing stress-induced apoptosis in *atg7*$^{-/-}$ and *FIP200*$^{-/-}$ mice [79, 81]. And FIP200 (200 kDa focal adhesion kinase family interacting protein) has been shown to be essential not only for the induction of autophagy but also for the maintenance and function of HSCs in vivo [82, 83]. These results indicate that autophagy can maintain the health of HSCs and play a crucial role in the normal development of the body.

#### 1.3.2.2 The Effect of Autophagy on the Regeneration of Hematopoietic Stem Cells

Self-renewal of HSCs maintains their own survival and expansion to ensure that the hematopoietic process can be carried out throughout the body's life cycle. But the molecular mechanisms of self-renewal of HSCs are not well understood. Interestingly, current research has suggested that autophagy affects the regeneration of HSCs.

The majority of HSCs studies have suggested that Wnt signaling pathway plays an important role in maintaining the ability of HSCs to self-renew. The activation of mTOR by Wnt signaling pathway suppresses the process of autophagy [84]. Existing research has suggested that BMI1, a proto-oncogene of multiple combs, plays a central role in regulating the self-renewal of bone marrow HSCs [85]. In the *bmi1*$^{-/-}$ mouse model, the number of bone marrow HSCs decreased significantly and the secondary colony formation was significantly reduced in vitro. And self-renewed HSCs could not be detected. The result indicates that *bmi1*$^{-/-}$ HSCs lack the capacity to self-renew [86].

In addition, there are evidences have suggested that the accumulation of the self-renewal capacity of human HSCs are limited by the accumulation of damaged DNA [87]. Numerous studies have demonstrated that mitophagy is a powerful means of cellular to regulate intracellular ROS levels and DNA damage accumulation [88]. The use of antioxidants or rapamycin restored HSCs self-renewal at high ROS levels [89, 90]. These studies suggest that autophagy may regulate the self-renewal of HSCs by regulating intracellular ROS levels.

By regulating mTOR and ROS levels, the state of HSCs can be altered. Although HSCs are permanently in quiescence, there is unavoidable need repair and defense mechanisms to against cell damage caused by long-term silencing. Therefore, autophagy, as a protector that repairs cell damage, not only maintains the quiescence of HSCs but also promotes self-renewal of HSCs in order to achieve regulation of HSCs.

#### 1.3.2.3 The Effect of Autophagy on the Proliferation and Differentiation of Hematopoietic Stem Cells

All mature blood cells are derived from the directional differentiation of HSCs. In addition, autophagy plays a role in clearing organelles such as mitochondria during the differentiation of MSCs to erythrocytes [91]. Although autophagy plays an important role in hematopoietic differentiation has been widely reported, the specific regulatory mechanism is still unknown.

It has been reported that Notch signaling pathway takes part in the autonomic regulation of HSCs fate. There are studies have found that the level of ROS is gradually increased during differentiation of HSCs, and the inhibition of ROS can reduce differentiation of HSCs [92]. Studies suggest that the inhibition of ROS downregulates autophagy in HSCs and progenitor cells (HSPCs) of wild-type mice, but Notch signaling pathway and its downstream are activated. In contrast, ROS suppression did not alter myeloid differentiation in HSPCs of autophagy complete deficient mice. And mitochondrial ROS, which was produced during metabolism, triggered autophagy to downregulate Notch pathway and then promoted the differentiation of HSCs. Therefore, autophagy can regulate the differentiation of HSCs to maintain hematopoiesis by directly targeting Notch signaling pathway [93, 94]. Furthermore, Gata-1 is an important transcription factor that regulates erythroid differentiation of HSCs. Recently, it was found that Gata-1 activated the expression of autophagy-related genes such as *LC3b*, *ATG*, *MAP1-LC3A*, *GABARAP*, and *GABARAPL1*. With the maturation of erythroid differentiation of HSCs, the organelles were cleared by autophagy, which was activated by upregulating the expression of Gata-1 and autophagy-related gene, and then erythrocytes without organelles were formed [33].

The above available evidences indicate that autophagy plays an important role in cell remodeling and pluripotency of HSCs (Fig. 1.2).

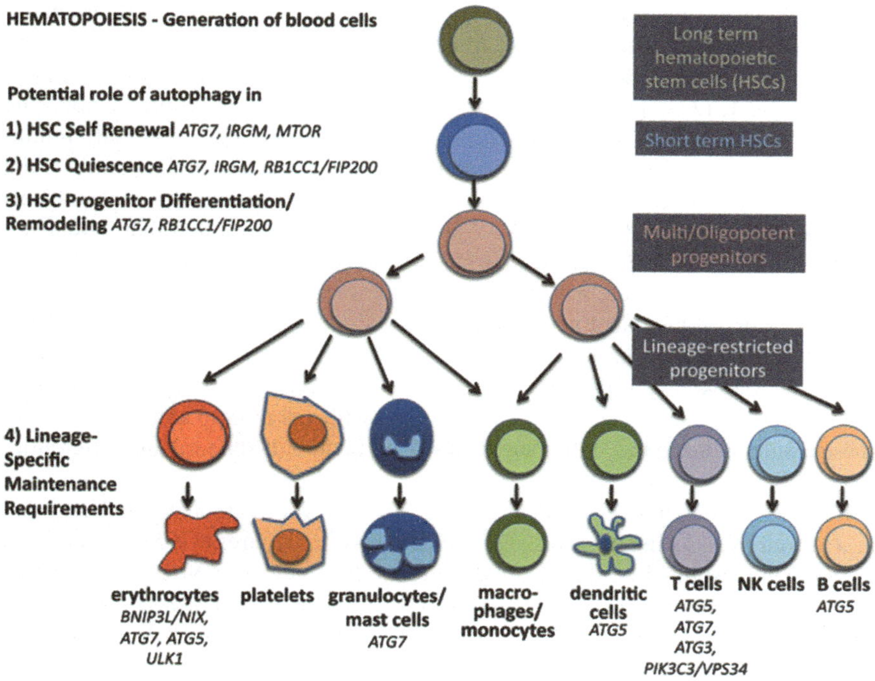


**Fig. 1.2** Effect of autophagy on hematopoietic stem cells. Hematopoietic stem cells produce different types of blood cells through hematopoiesis. Autophagy is thought to play an important role in self-renewal, quiescence, and differentiation in HSCs to make long-term HSCs develop into short-term HSCs, multipotent/oligopotent progenitors, and lineage-restricted progenitors, which generate differentiated blood cells including erythrocytes, platelets, granulocytes, macrophages, dendritic cells, and the lymphocytes T, B, and NK cells. And it is hypothesized that autophagy help HSCs maintain multipotency and remodeling. Permission obtained from TAYLOR & FRANCIS LICENCE © Guan, J.L., et al. Autophagy, 2013. 9(6): p. 830–49

### 1.3.3 Autophagy and Neural Stem Cells

Survival, apoptosis, proliferation and differentiation of NSCs are regulated by many factors. NSCs display a variety of changes under different external stimuli. There is growing evidence that autophagy is involved in the apoptosis and differentiation of NSCs and plays a very important role in brain development, synaptic remodeling, neurodegenerative diseases, and tumorigenesis [95–99]. For instance, Zeng Mei et al. demonstrated that the process of autophagy was activated in neuroblastoma N2a cells that were differentiated induced by retinoic acid. And the treatment with the inhibitors of autophagy such as 3-MA and LY294002 or knock down *beclin 1* via siRNA markedly inhibited the process of cells differentiation [100]. Moreover Zeng Mei et al. also suggested that cell differentiation was accompanied by decreasing the activity of mTOR that control cell growth and regulate autophagy [100]. And Li et al. have showed that the generation of newborn neurons were impaired during the protein of Eva1a was depleted. Conversely, overexpression of Eva1a enhanced newborn neuron generation and maturation. It is suggested that Eva1a (autophagy-related protein) regulates NSC self-renewal and differentiation by modulating autophagy [101]. Meanwhile Vazquez et al. have suggested that the expression of Atg7, Beclin 1, LC3, and Ambra1 is significantly increased in cultured embryonic mouse olfactory bulb-derived NSCs during the initial period of neuronal differentiation [102]. In addition, a number of studies have shown that autophagy affects the proliferation and differentiation of NSCs primarily through regulation of intracellular oxidative stress [103, 104]. As Wang CF et al. have implied that autophagy function is lost in subventricular zone (SVZ) and dentate gyrus cells on *FIP200*$^{-/-}$ and *p53*$^{-/-}$ knockout mice, but subsequent experiments demonstrate that p53 does not affect FIP200-mediated neuronal differentiation, whereas the removal of ROS effectively attenuate disorders of neuronal regeneration and differentiation induced by FIP200 knockdown [105]. In addition to *beclin1*, the expressions of autophagy-related genes *atg3*, *atg5*, and *atg7* are decreased with age in vitro and in vivo [106]. These results indicate that autophagy is involved in the self-renewal and differentiation of NSCs.

Further research has shown that the protective effect of autophagy on NSCs can be achieved by reducing the level of intracellular oxidative stress [105]. Knockdown FIP200 gene resulted in a significant decrease in the number of NSCs, a significant increase of the number of apoptotic cells in SVZ and subgranular zone (SGZ), and a significant increase in ROS levels in NSCs, while scavenging ROS could effectively reduce apoptosis of NSCs caused by FIP200 knockout in SVZ and SGZ [105]. Consistent with the above view, the deletion of Prdm16 led to changes in an increased ROS production and increased cell death so that abrogated NSCs self-renewal, which can be rescued defects in NSCs function treating with the antioxidant *N*-acetyl cysteine [107]. In addition, Paik JH et al. have found that inactivation of FoxO leads to defective self-renewal and differentiation of NSCs accompanied with increased ROS [108]. However, there are some disagreements that ROS levels could influence their self-renewal properties in NSCs. For example, Le Belle et al. have implied that the self-renewal and neurogenesis capabilities of NSCs depend on high levels of ROS in the mouse SVZ [103]. The authors proposed that NSCs might maintain high ROS levels during highly proliferative stages of development, and lower levels during quiescence, suggesting a mechanism for antioxidant regulation [71]. Therefore, the kind of impact autophagy induced by ROS has on NSCs needs further study.

### *1.3.4 Autophagy and Mesenchymal Stem Cells*

MSCs have multidirectional differentiation potential and can differentiate into adipose, bone, cartilage and muscle tissues. Although the original MSCs are isolated from human bone marrow, up to now, it has been isolated from many other adult tissues such as muscles, adipose tissues, kidney, pancreas, brain, and liver [44]. Currently, the role of autophagy in MSCs is gradually clear. The initial knowledge about autophagy and MSCs is derived from a very few studies using bone marrow MSCs (BM-MSCs). Youjin Lee et al. have found that the expression of LC3 II is more rapidly increased in chorionic plate-derived-MSCs than in BM-MSCs under hypoxia condition, and autophagy is activated to increase SCF/c-kit pathway facilitating the self-renewal of CP-MSCs and balance cell survival and death. Additionally, Sedigheh Molaei et al. have indicated that the induction of autophagy gives rise to cell death under stressful conditions such as hypoxia, serum deprivation and oxidative stress, while the inhibition of autophagy, by downregulating the autophagy gene, ATG7, promoted MSCs withstand the stress conditions [109]. Besides this above knowledge in MSCs, there is emerging evidence for the important role of autophagy in MSCs-derived cells including adipocytes, chondrocytes, and osteoblasts/osteocytes [33]. But at current stage of the studies, it has found that autophagy can regulate the proliferation, differentiation, apoptosis, and survival of MSCs. For instance, Lisa Oliver et al. have demonstrated that human bone marrow MSCs exhibit a high level of autophagy, whereas the survival and differentiation of human bone marrow MSCs are suppressed, in which these cells differentiate into osteoblasts, during the process that autophagy is inhibited by knockdown Bcl-xL [110]. Bo Li et al. have indicated that the induction of autophagy in BMSCs decreases its S-phase population, and induces its differentiation into neurons [111]. Moreover, Wan Y et al. have found that activation of autophagy induced by rapymycin significantly improves osteogenic differentiation [112]. An important physiological function of autophagy is to maintain cell viability and protect cells from apoptosis under starvation conditions. Moreover, there is a close relationship between autophagy and apoptosis [45, 113, 114]. Hence, a previous study has shown that atorvastatin decreases the activation of apoptosis through enhancing the autophagic activity under hypoxia/serum deprivation condition [115]. Instead, Cecilia G. Sanchez et al. have suggested that serum-deprived MSCs (SD-MSCs) exhibit higher cellularity, decrease apoptosis, and decrease differentiation during coculturing breast cancer cells in medium. It has been suggested that solid tumors occur in nutrient-deprived conditions, and MSCs utilize autophagy for survival and secrete antiapoptotic factors, which can promote solid tumor survival and growth [116]. Although above results are not the same, they all show that autophagy plays a very important role in the physiological regulation of MSCs. In summary, autophagy is crucial for the fate and function of MSCs.

### *1.3.5 Autophagy and Cancer Stem Cells*

Cancer stem cells (CSCs) not only have the ability to self-renew and differentiate as same as other SCs, such as ESCs, MSCs and NSCs, but also are a subpopulation of cancer cells [117–119]. Recent studies have suggested that autophagy plays an important role in the occurrence and development of tumors, and the role of autophagy is not the same in different stages of tumor growth. In a word, autophagy inhibits tumor cell formation during the initial stages of tumor formation, whereas promotes tumor cell survival after tumor maturation. Herein, we discussed the role of autophagy in CSCs.

On the one hand, some studies have shown that the induction of autophagy is conducive to clear CSCs in body. In tumor tissue, CSCs are often exposed to hypoxia and undernutrition, which promotes the survival of CSCs via activating the process of autophagy that can promote the recycling of intracellular substances [120]. After inhibition of autophagy, breast cancer stem cells and hepatocellular carcinoma stem cells were more prone to apoptosis, and reduced the ability of CSCs to form tumors [120]. Then Liang Xiao Huan et al. have found that endogenous Beclin1 protein expression was frequently low in human breast epithelial carcinoma cell lines and tissue, and up-regulated *beclin1* gen could inhibit MCF7 cellular proliferation, in vitro clonigenicity and tumorigenesis in nude mice [121].

On the other hand, autophagy can also promote tumor progression, such as the spread and metastasis of tumor cells. It was shown that autophagy supported the survival of breast malignant precursor cells but the generation of breast ductal carcinoma and xenograft tumor formation was abrogated during the process of autophagy inhibition by chloroquine, the inhibitor of autophagy [122]. Similarly, Cufi et al. and Maycotte et al. have proved that loss of the proteins of ATG7, ATG12 or ATG8/LC3 receded the growth of $CD44^{+}CD24^{-/low}$ breast cancer stem cells in vitro [123, 124]. At the same time it is interesting to note that inhibited-autophagy can reduce the secretion of IL6, which plays a key cytokine in the maintenance of CSCs, in $CD44^{+}CD24^{-}$ breast CSCs [123]. Therefore, Galluzzi et al. have considered that active autophagy serves as a tumor-suppressing mechanism or tumor-promoting mechanism depending on the type of cancer and the stage of development [125].

As the role of autophagy on CSCs has two sides (Figs. 1.3 and 1.4), we need activate or inhibit the process of autophagy to achieve the purpose of treatment of cancer according to the specific situation.

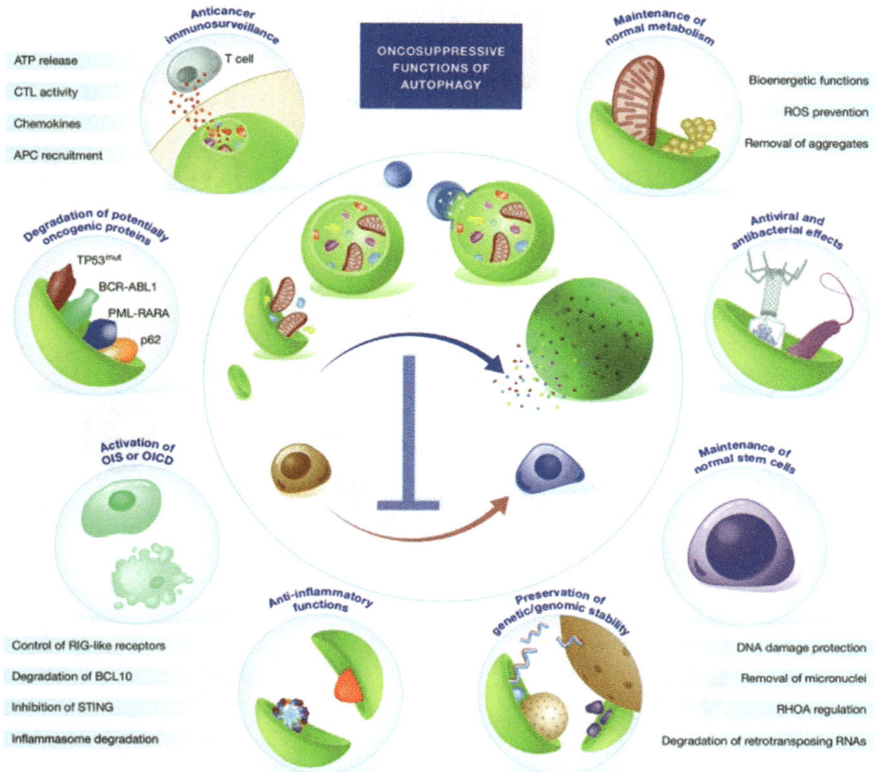


**Fig. 1.3** Oncosuppressive functions of autophagy. Autophagy has been proved to inhibit tumor metastasis through different mechanisms, including (1) the maintenance of normal metabolism; (2) the effects of antiviral and antibacterial; (3) the maintenance of normal stem cell; (4) the preservation of genetic/genomic stability; (5) the function of anti-inflammatory; (6) the activation of oncogene-induced senescence (OIS) or oncogene-induced cell death (OICD); (7) the degradation of potentially oncogenic proteins; (8) anticancer immunosurveillance. *ABL1* ABL proto-oncogene 1, *APC* antigen-presenting cell, *BCL10* B-cell CLL/lymphoma 10, *BCR* breakpoint cluster region, *CTL* cytotoxic T lymphocyte, *TP53mut* mutant tumor protein p53, *PML* promyelocytic leukemia, *RARA* retinoic acid receptor, alpha, *RHOA* ras homolog family member. Permission obtained from JOHN WILEY & SONS LICENCE © Galluzzi, L., et al. Embo Journal, 2015. 34(7): p. 856–880

## 1.4 Progress of Autophagy and Stem Cells in Different Diseases

With the expansion of aging population, neurodegenerative diseases and tumor have become urgent problems to be solved. And as technology advances, investigators have found that stem cells can be used to improve neurodegenerative diseases and cancer.

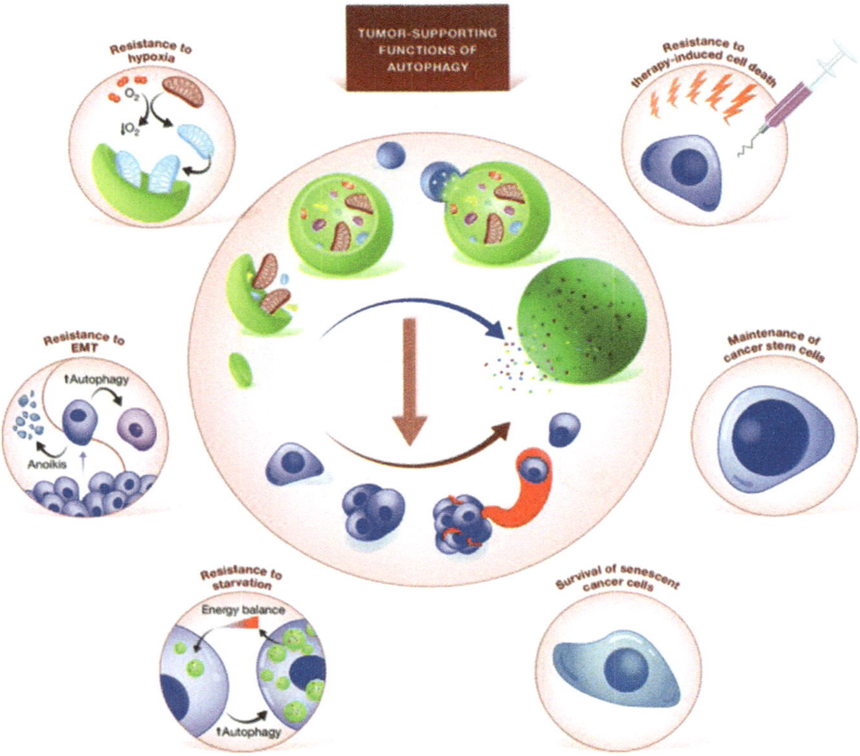


**Fig. 1.4** Tumor-supporting functions of autophagy. Once the tumor has been formed, autophagy is believed to promote tumor progression by different mechanisms including: (1) the resistance of therapy-induced cell death or EMT; (2) the maintenance of cancer stem cells; (3) sustain the survival of senescent cancer cells; (4) the resistance of endogenous bad conditions such as starvation or hypoxia. *EMT* epithelial-to-mesenchymal transition. Permission obtained from JOHN WILEY AND SONS LICENSE © Galluzzi, L., et al.-Embo Journal, 2015. 34(7): p. 856–880

On the one hand, more and more research works focus on the treatment of neurodegenerative diseases by stem cell transplantation. For example, Wen Li et al. have suggested that the activation of autophagy in huMSCs leads to the appearance of functions including migration, differentiation, and antiapoptosis, and the promotion of neurogenesis and synapse formation in the APP/PS1 transgenic mice [45]. The result hints that the therapeutic effect of huMSCs on neurodegenerative diseases can be improved by increasing the level of autophagy-utilizing medicines or small molecule compounds. Furthermore, Wei Yue et al. have shown that the induction of embryonic stem cells into cholinergic precursor cells transplanted into AD model mice improved the cognitive function of mice because ESC-derived BFCNs can replace and partial reconstruct the damaged cholinergic neuron circuitry of AD mice [126]. And Qing Ying Wu et al. have shown that bone marrow MSCs can be differentiated into cholinergic neurons after they are transplanted into the brain of mice, thereby reducing the number of senile plaques and improving the learning and memory of AD rats [127].

On the other hand, one of the current methods that can test biochemical markers of neurodegenerative diseases is autopsy. However, this approach is inherently limited due to the impossible of studying patient neurons prior to degeneration. Fortunately, the generation of patient-specific induced pluripotent stem (iPS) cells offers a unique opportunity to overcome these obstacles. By expanding and differentiating iPS cells, it is possible to generate large numbers of functional neurons in vitro, which can then be used to study the disease of the donating patient [128].

In cancer, some studies have shown that induced-autophagy drugs may be beneficial in preventing the growth and progression of tumor cells, and inhibited-autophagy drugs may improve the efficacy of anticancer therapies in patients with metastatic cancer [129, 130].

Whether it is for improving neurodegenerative diseases or treating cancer, we have to understand that much work remains to be done before developing an approach for altering NSC-related or CSC-related autophagy in a clinical setting.

## 1.5 Application Prospects and Summary

More and more studies have uncovered the role of autophagy, a highly conserved mechanism of intracellular clearance, in stem cells. This stage of research shows that the characteristics of self-renewal, pluripotency, and proliferation of stem cells are closely related with autophagy. These results suggest that we can control, visualize, and create stem cells in the future by regulating the process of autophagy.

## References

1. Klionsky DJ. Autophagy revisited: a conversation with Christian de Duve. Autophagy. 2008;4(6):740–3.
2. Kim KH, Lee MS. Autophagy--a key player in cellular and body metabolism. Nat Rev Endocrinol. 2014;10(6):322.
3. Levine B, Yuan J. Autophagy in cell death: an innocent convict? J Clin Investig. 2005;115(10):2679–88.
4. Komatsu M, et al. Impairment of starvation-induced and constitutive autophagy in Atg7-deficient mice. J Cell Biol. 2005;169(3):425.
5. Cuervo AM, Wong E. Chaperone-mediated autophagy: roles in disease and aging. Cell Res. 2014;24(1):92–104.
6. Galluzzi L, et al. Pharmacological modulation of autophagy: therapeutic potential and persisting obstacles. Nat Rev Drug Discov. 2017;16(7):487.
7. Zhang J, et al. Mechanisms of autophagy and relevant small-molecule compounds for targeted cancer therapy. Cell Mol Life Sci. 2018;75:1803.
8. Mizushima N, Yoshimori T, Ohsumi Y. The role of atg proteins in autophagosome formation. Annu Rev Cell Dev Biol. 2011;27(1):107.

9. Ke B, et al. Targeting programmed cell death using small-molecule compounds to improve potential cancer therapy. Med Res Rev. 2016;36(6):983–1035.
10. Hosokawa N, et al. Nutrient-dependent mTORC1 association with the ULK1-Atg13-FIP200 complex required for autophagy. Mol Biol Cell. 2009;20(7):1981.
11. Chang HJ, et al. ULK-Atg13-FIP200 complexes mediate mTOR signaling to the autophagy machinery. Mol Biol Cell. 2009;20(7):1992.
12. Chang YY, Neufeld TP. An Atg1/Atg13 complex with multiple roles in TOR-mediated autophagy regulation. Mol Biol Cell. 2009;20(7):2004.
13. Itakura E, et al. Beclin 1 forms two distinct phosphatidylinositol 3-kinase complexes with mammalian Atg14 and UVRAG. Mol Biol Cell. 2008;19(12):5360–72.
14. Di BS, et al. The dynamic interaction of AMBRA1 with the dynein motor complex regulates mammalian autophagy. J Cell Biol. 2011;191(1):155–68.
15. Axe EL, et al. Autophagosome formation from membrane compartments enriched in phosphatidylinositol 3-phosphate and dynamically connected to the endoplasmic reticulum. J Cell Biol. 2008;182(4):685.
16. Polson HE, et al. Mammalian Atg18 (WIPI2) localizes to omegasome-anchored phagophores and positively regulates LC3 lipidation. Autophagy. 2010;6(4):506.
17. Mizushima N, et al. A new protein conjugation system in human. The counterpart of the yeast Apg12p conjugation system essential for autophagy. J Biol Chem. 1998;273(51):33889–92.
18. Mizushima N, et al. A protein conjugation system essential for autophagy. Nature. 1998;395(6700):395–8.
19. Kuma A, et al. Formation of the approximately 350-kDa Apg12-Apg5.Apg16 multimeric complex, mediated by Apg16 oligomerization, is essential for autophagy in yeast. J Biol Chem. 2002;277(21):18619.
20. Kabeya Y, et al. LC3, a mammalian homologue of yeast Apg8p, is localized in autophagosome membranes after processing. EMBO J. 2000;19(21):5720–8.
21. Tanida I, et al. HsAtg4B/HsApg4B/autophagin-1 cleaves the carboxyl termini of three human Atg8 homologues and delipidates microtubule-associated protein light chain 3- and GABAA receptor-associated protein-phospholipid conjugates. J Biol Chem. 2004;279(35):36268–76.
22. Ichimura Y, et al. A ubiquitin-like system mediates protein lipidation. Nature. 2000;408(6811):488.
23. Klionsky DJ, Emr SD. Autophagy as a regulated pathway of cellular degradation. Science. 2000;290(5497):1717–21.
24. Yu L, et al. Termination of autophagy and reformation of lysosomes regulated by mTOR. Nature. 2010;465(7300):942–6.
25. Shen R, et al. The function and regulation of BMP6 in various kinds of stem cells. Curr Pharm Des. 2015;21(25):3634–43.
26. He S, Nakada D, Morrison SJ. Mechanisms of stem cell self-renewal. Annu Rev Cell Dev Biol. 2009;25(25):377.
27. Bongso A, Lee EH. Stem cells: their definition, classification and sources. Stem Cells. London: World Scientific Publishing Co. Pte. Ltd.; 1935. p. 1–13.
28. Dave M, et al. Mesenchymal stem cell therapy for inflammatory bowel disease: a systematic review and meta-analysis. Inflamm Bowel Dis. 2015;21(11):2696.
29. Hsieh MM, et al. Allogeneic hematopoietic stem-cell transplantation for sickle cell disease. N Engl J Med. 2009;361(24):2309.
30. Laird DJ, Andrian UHV, Wagers AJ. Stem cell trafficking in tissue development, growth, and disease. Cell. 2008;132(4):612.
31. Pera MF, Reubinoff B, Trounson A. Human embryonic stem cells. J Cell Sci. 2000;113(Pt 1):5.
32. Pera MF, Dottori M. Stem cells and their developmental potential. Singapore: World Scientific Publishing; 2015. p. 55–70.
33. Guan JL, et al. Autophagy in stem cells. Autophagy. 2013;9(6):830–49.

34. Evans MJ, Kaufman MH. Establishment in culture of pluripotential cells from mouse embryos. Nature. 1981;292(5819):154–6.
35. Martin GR. Isolation of a pluripotent cell line from early mouse embryos cultured in medium conditioned by teratocarcinoma stem cells. Proc Natl Acad Sci U S A. 1981;78(12):7634–8.
36. Bradley A, et al. Formation of germ-line chimaeras from embryo-derived teratocarcinoma cell lines. Nature. 1984;309(5965):255–6.
37. Ross EA, Anderson N, Micklem HS. Serial depletion and regeneration of the murine hematopoietic system. Implications for hematopoietic organization and the study of cellular aging. J Exp Med. 1982;155(2):432–44.
38. Orkin SH, Zon LI. Hematopoiesis: an evolving paradigm for stem cell biology. Cell. 2008;132(4):631.
39. Kumaravelu P, et al. Quantitative developmental anatomy of definitive haematopoietic stem cells/long-term repopulating units (HSC/RUs): role of the aorta-gonad-mesonephros (AGM) region and the yolk sac in colonisation of the mouse embryonic liver. Development. 2002;129(21):4891.
40. Rhodes KE, et al. The emergence of hematopoietic stem cells is initiated in the placental vasculature in the absence of circulation. Cell Stem Cell. 2008;2(3):252.
41. Pittenger MF, et al. Multilineage potential of adult human mesenchymal stem cells. Science. 1999;284(5411):143–7.
42. Majumdar MK, et al. Phenotypic and functional comparison of cultures of marrow-derived mesenchymal stem cells (MSCs) and stromal cells. J Cell Physiol. 1998;176(1):57–66.
43. Caplan AI. The mesengenic process. Clin Plast Surg. 1994;21(3):429.
44. Nombela-Arrieta C, Ritz J, Silberstein LE. The elusive nature and function of mesenchymal stem cells. Nat Rev Mol Cell Biol. 2011;12(2):126–31.
45. Li W, et al. Autophagy is required for human umbilical cord mesenchymal stem cells to improve spatial working memory in APP/PS1 transgenic mouse model. Stem Cell Res Ther. 2018;9(1):9.
46. Maitra B, et al. Human mesenchymal stem cells support unrelated donor hematopoietic stem cells and suppress T-cell activation. Bone Marrow Transplant. 2004;33(6):597–604.
47. Groszer M, et al. PTEN negatively regulates neural stem cell self-renewal by modulating G0-G1 cell cycle entry. Proc Natl Acad Sci U S A. 2006;103(1):111–6.
48. Jackson EL, et al. PDGFRα-Positive B cells are neural stem cells in the adult SVZ that form glioma-like growths in response to increased PDGF signaling. Neuron. 2006;51(2):187–99.
49. Hermanson O, Jepsen K, Rosenfeld MG. N-CoR controls differentiation of neural stem cells into astrocytes. Nature. 2002;419(6910):934–9.
50. Aguirre A, Rubio ME, Gallo V. Notch and EGFR pathway interaction regulates neural stem cell number and self-renewal. Nature. 2010;467(7313):323.
51. Gage FH. Mammalian neural stem cells. Science. 2000;287(5457):1433.
52. Phadwal K, Watson AS, Simon AK. Tightrope act: autophagy in stem cell renewal, differentiation, proliferation, and aging. Cell Mol Life Sci. 2013;70(1):89–103.
53. Vessoni AT, Muotri AR, Okamoto OK. Autophagy in stem cell maintenance and differentiation. Stem Cells Dev. 2012;21(4):513–20.
54. Tooze J, Davies HG. Cytolysomes in amphibian erythrocytes. J Cell Biol. 1965;24(1):146.
55. Kent G, et al. Autophagic vacuoles in human red cells. Am J Pathol. 1966;48(48):831–57.
56. Kundu M, et al. Ulk1 plays a critical role in the autophagic clearance of mitochondria and ribosomes during reticulocyte maturation. Blood. 2008;112(4):1493–502.
57. Thomson JA, et al. Embryonic stem cell lines derived from human blastocysts. Science. 1998;282(5391):1145.
58. Chen Y, Klionsky DJ. The regulation of autophagy – unanswered questions. J Cell Sci. 2011;124(Pt 2):161–70.
59. He C, Klionsky DJ. Regulation mechanisms and signaling pathways of autophagy. Annu Rev Genet. 2009;43(1):67.

60. Mizushima N, et al. Dissection of autophagosome formation using Apg5-deficient mouse embryonic stem cells. J Cell Biol. 2001;152(4):657–68.
61. Tsukamoto S, et al. Autophagy is essential for preimplantation development of mouse embryos. Science. 2008;321(5885):117–20.
62. Wang S, et al. Transient activation of autophagy via Sox2-mediated suppression of mTOR is an important early step in reprogramming to pluripotency. Cell Stem Cell. 2013;13(5):617.
63. Mizushima N. Aβ generation in autophagic vacuoles. J Cell Biol. 2005;171(1):15.
64. Cho YH, et al. Autophagy regulates homeostasis of pluripotency-associated proteins in hESCs. Stem Cells. 2014;32(2):424–35.
65. Tra T, et al. Autophagy in human embryonic stem cells. PLoS One. 2011;6(11):e27485.
66. Qu X, et al. Autophagy gene-dependent clearance of apoptotic cells during embryonic development. Cell. 2007;128(5):931–46.
67. Saitoh T, et al. Atg9a controls dsDNA-driven dynamic translocation of STING and the innate immune response. Proc Natl Acad Sci U S A. 2009;106(49):20842.
68. Sou YS, et al. The Atg8 conjugation system is indispensable for proper development of autophagic isolation membranes in mice. Mol Biol Cell. 2008;19(11):4762.
69. Kuma A, et al. The role of autophagy during the early neonatal starvation period. Nature. 2004;432(7020):1032–6.
70. Levine B. Autophagy in mammalian development and differentiation. Nat Cell Biol. 2010;12(9):823.
71. Boya P, Codogno P, Rodriguez-Muela N. Autophagy in stem cells: repair, remodelling and metabolic reprogramming. Development. 2018;145(4):pii: dev146506.
72. Yue Z, et al. Beclin 1, an autophagy gene essential for early embryonic development, is a haploinsufficient tumor suppressor. Proc Natl Acad Sci U S A. 2003;100(25):15077–82.
73. Chen S, et al. Distinct roles of autophagy-dependent and -independent functions of FIP200 revealed by generation and analysis of a mutant knock-in mouse model. Genes Dev. 2016;30(7):856–69.
74. Kaizuka T, Mizushima N. Atg13 is essential for autophagy and cardiac development in mice. Mol Cell Biol. 2016;36(4):585–95.
75. Kuo TC, et al. Midbody accumulation through evasion of autophagy contributes to cellular reprogramming and tumorigenicity. Nat Cell Biol. 2011;13(10):1214–23.
76. Schink KO, et al. Cell differentiation: midbody remnants — junk or fate factors? Curr Biol. 2011;21(23):958–60.
77. Isakson P, et al. TRAF6 mediates ubiquitination of KIF23/MKLP1 and is required for midbody ring degradation by selective autophagy. Autophagy. 2013;9(12):1955.
78. Mandell MA, et al. TRIM17 contributes to autophagy of midbodies while actively sparing other targets from degradation. J Cell Sci. 2016;129(19):3562.
79. Mortensen M, et al. The autophagy protein Atg7 is essential for hematopoietic stem cell maintenance. J Exp Med. 2011;208(3):455.
80. Warr MR, et al. FOXO3A directs a protective autophagy program in haematopoietic stem cells. Nature. 2013;494(7437):323–7.
81. Liu F, Guan JL. FIP200, an essential component of mammalian autophagy is indispensible for fetal hematopoiesis. Autophagy. 2011;7(2):229.
82. Salemi S, et al. Autophagy is required for self-renewal and differentiation of adult human stem cells. Cell Res. 2012;22(2):432.
83. Liu F, et al. FIP200 is required for the cell-autonomous maintenance of fetal hematopoietic stem cells. Blood. 2010;116(23):4806.
84. Kang Y, et al. Autophagy driven by a master regulator of hematopoiesis. Mol Cell Biol. 2012;32(1):226–39.
85. Iwama A, et al. Enhanced self-renewal of hematopoietic stem cells mediated by the polycomb gene product Bmi-1. Immunity. 2004;21(6):843.
86. Park IK, et al. Bmi-1 is required for maintenance of adult self-renewing haematopoietic stem cells. Nature. 2003;423(6937):302.

87. Yahata T, et al. Accumulation of oxidative DNA damage restricts the self-renewal capacity of human hematopoietic stem cells. Blood. 2011;118(11):2941.
88. Joshi A, Kundu M. Mitophagy in hematopoietic stem cells: the case for exploration. Autophagy. 2013;9(11):1737.
89. Chong C, et al. TSC–mTOR maintains quiescence and function of hematopoietic stem cells by repressing mitochondrial biogenesis and reactive oxygen species. J Exp Med. 2008;205(10):2397.
90. Chen C, et al. mTOR regulation and therapeutic rejuvenation of aging hematopoietic stem cells. Sci Signal. 2009;2(98):ra75.
91. Mortensen M, et al. Loss of autophagy in erythroid cells leads to defective removal of mitochondria and severe anemia in vivo. Autophagy. 2010;107(3):832–7.
92. Yan C, et al. ROS functions as an upstream trigger for autophagy to drive hematopoietic stem cell differentiation. Hematology. 2016;21(10):1.
93. Cao Y, et al. Autophagy sustains hematopoiesis through targeting notch. Stem Cells Dev. 2015;24(22):2660.
94. Cao Y, et al. Hierarchal autophagic divergence of hematopoietic system. J Biol Chem. 2015;290(38):23050–63.
95. Yuan X, et al. Mitochondrial apoptosis and autophagy in the process of adipose-derived stromal cell differentiation into astrocytes. Cell Biol Int. 2016;40(2):156–65.
96. Denis JA, et al. mTOR-dependent proliferation defect in human ES-derived neural stem cells affected by myotonic dystrophy type 1. J Cell Sci. 2013;126(8):1763–72.
97. Shang YC, et al. Prevention of β-amyloid degeneration of microglia by erythropoietin depends on Wnt1, the PI 3-K/mTOR pathway, Bad, and Bcl-xL. Aging. 2012;4(3):187–201.
98. Walls KC, et al. Lysosome dysfunction triggers Atg7-dependent neural apoptosis. J Biol Chem. 2010;285(14):10497–507.
99. Jiao Q. et al. [An overview on autophagy in neural stem cells]. Sheng li xue bao : [Acta Physiologica Sinica]. 2016; 68(5): 649.
100. Zeng M, Zhou JN. Roles of autophagy and mTOR signaling in neuronal differentiation of mouse neuroblastoma cells. Cell Signal. 2008;20(4):659.
101. Li M, et al. EVA1A/TMEM166 regulates embryonic neurogenesis by autophagy. Stem Cell Rep. 2016;6(3):396–410.
102. Vázquez P, et al. Atg5 and Ambra1 differentially modulate neurogenesis in neural stem cells. Autophagy. 2012;8(2):187.
103. Belle JEL, et al. Proliferative neural stem cells have high endogenous ROS levels that regulate self-renewal and neurogenesis in a PI3K/Akt-dependant manner. Cell Stem Cell. 2011;8(1):59.
104. Han X, et al. AMPKactivation protects cells from oxidative stress-induced senescence via autophagic flux restoration and intracellularNAD+elevation. Aging Cell. 2016;15(3):416–27.
105. Wang C, et al. FIP200 is required for maintenance and differentiation of postnatal neural stem cells. Nat Neurosci. 2013;16(5):532.
106. Zhu L, et al. Rejuvenation of MPTP-induced human neural precursor cell senescence by activating autophagy. Biochem Biophys Res Commun. 2015;464(2):526–33.
107. Chuikov S, et al. Prdm16 promotes stem cell maintenance in multiple tissues, partly by regulating oxidative stress. Nat Cell Biol. 2010;12(10):999.
108. Paik J, et al. FoxOs cooperatively regulate diverse pathways governing neural stem cell homeostasis. Cell Stem Cell. 2009;5(5):540–53.
109. Molaei S, et al. Down-regulation of the autophagy gene, ATG7, protects bone marrow-derived mesenchymal stem cells from stressful conditions. Blood Res. 2015;50(2):80–6.
110. Oliver L, et al. Basal autophagy decreased during the differentiation of human adult mesenchymal stem cells. Stem Cells Dev. 2012;21(15):2779.
111. Li B, et al. Role of autophagy on bone marrow mesenchymal stem-cell proliferation and differentiation into neurons. Mol Med Rep. 2016;13(2):1413–9.

112. Wan Y, et al. Autophagy promotes osteogenic differentiation of human bone marrow mesenchymal stem cell derived from osteoporotic vertebrae. Biochem Biophys Res Commun. 2017;488:46.
113. Karch J, et al. Autophagic cell death is dependent on lysosomal membrane permeability through Bax and Bak. Elife. 2017;6:e30543.
114. Musumeci G, et al. Biomarkers of chondrocyte apoptosis and autophagy in osteoarthritis. Int J Mol Sci. 2015;16(9):20560–75.
115. Zhang Q, et al. Autophagy activation: a novel mechanism of atorvastatin to protect mesenchymal stem cells from hypoxia and serum deprivation via AMP-activated protein kinase/mammalian target of rapamycin pathway. Stem Cells Dev. 2012;21(8):1321.
116. Sanchez CG, et al. Activation of autophagy in mesenchymal stem cells provides tumor stromal support. Carcinogenesis. 2011;32(7):964–72.
117. Clevers H. The cancer stem cell: premises, promises and challenges. Nat Med. 2011;17(3):313–9.
118. Magee JA, Piskounova E, Morrison SJ. Cancer stem cells: impact, heterogeneity, and uncertainty. Cancer Cell. 2012;21(3):283–96.
119. Visvader J, Lindeman G. cancer stem cells: current status and evolving complexities. Cell Stem Cell. 2012;10(6):717.
120. Gong C, et al. Beclin 1 and autophagy are required for the tumorigenicity of breast cancer stem-like/progenitor cells. Autophagy. 2012;8(12):1853–5.
121. Liang XH, et al. Induction of autophagy and inhibition of tumorigenesis by beclin 1. Nature. 1999;402(6762):672.
122. Espina V, et al. Malignant precursor cells pre-exist in human breast DCIS and require autophagy for survival. PLoS One. 2010;5(4):e10240.
123. Maycotte P, et al. Autophagy supports breast cancer stem cell maintenance by regulating IL6 secretion. Mol Cancer Res. 2015;13(4):651.
124. Cufã SL, et al. Autophagy positively regulates the CD44+CD24-/low breast cancer stem-like phenotype. Cell Cycle. 2011;10(22):3871–85.
125. Galluzzi L, et al. Autophagy in malignant transformation and cancer progression. EMBO J. 2015;34(7):856–80.
126. Yue W, et al. ESC-derived basal forebrain cholinergic neurons ameliorate the cognitive symptoms associated with Alzheimer's disease in mouse models. Stem Cell Rep. 2015;5(5):776.
127. Wu QY, et al. Bone marrow stromal cells of transgenic mice can improve the cognitive ability of an Alzheimer's disease rat model. Neurosci Lett. 2007;417(3):281–5.
128. Hargus G, et al. Human stem cell models of neurodegeneration: a novel approach to study mechanisms of disease development. Acta Neuropathol. 2014;127(2):151.
129. Dalby KN, et al. Targeting the prodeath and prosurvival functions of autophagy as novel therapeutic strategies in cancer. Autophagy. 2010;6(3):322–9.
130. Buchser WJ, et al. Cell-mediated autophagy promotes cancer cell survival. Cancer Res. 2012;72(12):2970–9.

# Chapter 2
# Fine-Tuning the Stem Cell Fate by Autophagy

**Shalmoli Bhattacharyya and Ajay Kumar**

**Abstract** A constant balance is required between the anabolic and catabolic process to maintain cellular homeostasis. The cellular processing of cytoplasmic components by autophagy is the major pathway for intracellular degradation and recycling. Autophagy plays an important role in maintaining cellular homeostasis and tissue remodeling during normal development. Basal level of autophagy is prevalent in most tissues and it adds to the normal turnover of cytoplasmic components in the cell. Autophagy is also associated with health and longevity of dividing and differentiated cells. Dysregulation of autophagic pathways have been linked with the pathogenesis of diseases like cancer and various neurodegenerative disorders. The stem cells are a unique population of cells in the body having a high longevity and differentiation ability; hence autophagy is predicted to play a crucial role in maintenance of cellular homeostasis of these cells. Extensive information elucidating the function of autophagy in somatic cells is available but in contrast, the implication of autophagy in maintenance as well as differentiation of stem cells is being revealed recently. In this chapter, we discuss the recent updates in our knowledge of stem cell differentiation, quiescence, and the role of autophagy in their regulation.

## Abbreviations

| | |
|---|---|
| AKT | Serine/threonine kinase 3 |
| APL | Acute promyelocytic leukemia |
| ATG | Autophagy-related gene |
| bFGF | Basic fibroblast growth factor |
| BMMSC | Bone marrow-derived mesenchymal stem cells |
| CSC | Cancer stem cells |
| EGF | Epidermal growth factor |
| ESC | Embryonic stem cells |
| FIP200 | FAK-family interacting protein 200 |

S. Bhattacharyya (✉) · A. Kumar
Department of Biophysics, PGIMER, Chandigarh, India



K. Turksen (ed.), *Autophagy in Health and Disease*, Stem Cell Biology and Regenerative Medicine, https://doi.org/10.1007/978-3-319-98146-8_2

| | |
|---|---|
| HGF | Hepatocyte growth factor |
| HSC | Hematopoietic stem cells |
| IBMX | Isobutyl methyl xanthine |
| LC3 | Light chain 3 |
| MAPK | Mitogen-activated protein kinase |
| OSM | Oncostatin M |
| PDGF | Platelet-derived growth factor |
| PI3 | Phosphatidylinositol 3-kinase |
| PPAR | Peroxisome proliferator activator receptor |
| ROS | Reactive oxygen species |
| SC | Stem cell |
| STAT3 | Signal transducer and activator of transcription |

## 2.1 Stem Cell Differentiation

There are three important prerequisites for the practical application of stem cells for therapeutic purposes in regenerative medicine—first, the directed differentiation of stem cells to a desired phenotype; second, minimizing cell death and maximizing stem cell (SC) endurance after transplantation; and third, minimizing undifferentiated cells so that the risk of teratoma formation can be avoided. Apart from maintaining the stem cell fate, certain environmental cues can control the differentiation of stem cell to different lineages. Stem cell differentiation needs extensive exploration in context of the various networking pathways which comprehends more than just stem cell fate control. The property of differentiation makes stem cells a highly valuable tool for regenerative medicine. With the advent of new technologies and differentiation media, stem cells now can be robustly differentiated into cell lineages of all three germ layers with great ease and efficiency. This explains the derivation of required cell types from stem cells which become nonfunctional during injury/damage or aging; for example, in acute myocardial infarction, stem cells have been shown to improve overall cardiac functions including ventricular function, reduced infarct size, and enhanced angiogenesis in a rat model [1]. Stem cells have been also shown to provide rescuing effect in spinal cord injury model by differentiating into motor neurons and interacting with trigeminal neurons [2]. However, stem cells can show varying tendency to differentiate into different lineages based on the tissue from which they have originated, the propensity of these cells for a particular lineage has not been explored in detail. Our latest studies indicate that stem cells have an inherent propensity to differentiate into cells of different lineage based on their tissue of origin [3, 4]. The clinical application of stem cells derived from different sources may be guided by the inherent tendency of these cells to differentiate into a particular lineage which in turn may be dependent on various extrinsic and intrinsic factors. These factors should be considered while designing cell specific clinical trials in which the differentiation efficiency of chosen stem cell into cells of desired lineage can greatly affect the clinical trial outcome.

A wide array of agents which involve various cytokines, growth and transcription factors, and different chemical agents has been reported to induce differentiation in stem cells [5]. These factors offer a repertoire of molecules with wide choice to direct stem cell differentiation in desired fate for specific applications. A plethora of studies have shown that stem cells can be induced to differentiate by an alteration of niche after supplying various extrinsic and intrinsic cues [6]. The molecular mechanisms which control stem cell differentiation are slowly being unraveled. In monolayer cultures, stem cell differentiation is generally achieved by application of various chemicals like EGF, FGF, PDGF, dexamethasone, insulin, oxytocin, erythropoietin, and retinoic acid [5]. The differentiation protocols for stem cell differentiation varies among laboratories of the world but few very common chemicals are used for directed differentiation of stem cells into desired cell lineages [5]. We describe here few commonly used chemicals and their roles in differentiation into cells of mesodermal (osteocytes, adipocytes), ectodermal (neurons), and endodermal (hepatocytes) lineages.

Osteogenic differentiation of stem cells is generally carried out by application of dexamethasone, monopotassium phosphate ($KH_2PO_4$), and β-glycerophosphate. The low-dose dexamethasone is added to preserve the stem-like phenotype of hMSCs during repeated passaging as confirmed by the expression of stemness genes and multilineage differentiation potential at low doses. On the other hand, higher dose of dexamethasone was reported to promote the differentiation into osteocyte/adipocyte lineages [7]. In the past years, interest has deeply grown in adipogenesis with a particular emphasis on cross talk between extracellular signals and cascade of transcriptional factors which can regulate adipogenic differentiation [8]. Many different signals contribute to commitment of stem cells toward adipogenic lineage. Different chemical combinations including the use of dexamethasone, IBMX, and insulin are routinely used for adipogenic differentiation [9]; however, vast inconsistency exists between published protocols. Insulin is used primarily to increase the proliferation and differentiation of preadipocytes [10]. At higher concentration, it mimics insulin-like growth factor (IGF) and leads to activation of MAPK pathway [11, 12]. Dexamethasone is a cytokine which mainly reverses inflammation and can also lead to the development of osteogenic/adipogenic lineage in a cell-, time-, and concentration-dependent manner supporting an adipogenic concentration in MSC after prolonged exposure at higher concentration [13]. IBMX regulates PPAR-γ in combination with dexamethasone and promotes adipogenesis by enhancing expression of adipogenesis genes [13, 14]. Due to the increased number of debilitating neurodegenerative diseases like Alzheimer's and Parkinson's, the quest for a suitable stem cell source for obtaining sufficient number of neurons is ever increasing. Neural differentiation of stem cells can be achieved by a number of factors like retinoic acid, N2 supplement, basic fibroblast growth factor, and nerve growth factor [15–17]. Retinoic acid is involved in the switch from proliferation to differentiation during neuronal differentiation [18]. B-27 supplement is used for long-term survival of neurons, and in addition to bFGF, it favors the growth and attachment of mature neurons [19]. N2 supplement helps in the initial commitment and differentiation, survival, and expression of postmitotic neurons in culture. It favors the expression of neural specific genes, and it inhibits the growth of nonneuronal cells in the culture (or any undifferentiated cells) [20]. Sometimes G5 supplement

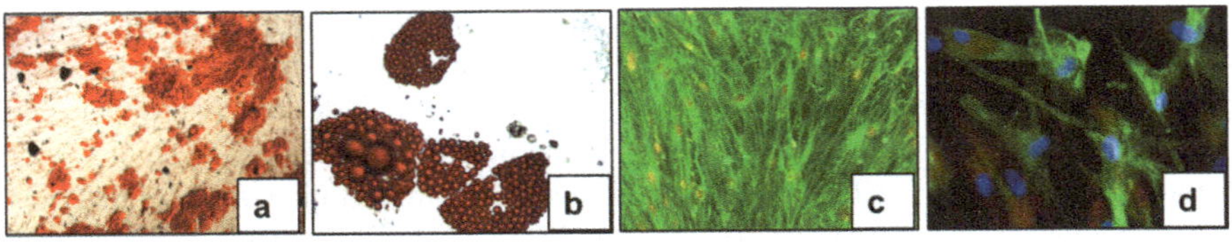


**Fig. 2.1** Directed differentiation of adult mesenchymal stem cells. (**a**) Differentiation of stem cells into osteocytes as demonstrated by uptake of alizarin red, calcium binding stain. (**b**) adipocyte differentiation indicated by uptake of Oil Red O stain by accumulated lipid droplets in differentiated mature adipocytes. (**c**) Neural differentiation of cells shown by immunofluorescence staining of axon cytoskeleton marker, Neurofilament antibody (green) using propidium iodide (red) as a nuclear stain. (**d**) Functional characterization of cells differentiated into hepatocytes by uptake of Low density Lipoproteins (LDL, red), LDL receptor (green) and DAPI as a nuclear stain

is also used in combination with N2 supplement to increase the efficacy of neural differentiation [20]. Hepatic differentiation of stem cells is a stepwise process including induction in first phase of differentiation and maturation in second one. Various chemicals like hepatocyte growth factor (HGF), oncostatin M (OSM), insulin, and transferrin can trigger the differentiation of stem cells into hepatocytes by stimulating various signaling pathways [21]. HGF can cause differentiation of oval cells in combination with other factors by a paracrine mechanism. It is also considered as a potent mitogen for hepatocytes in certain doses [22]. Many hepatic genes like tyrosine amino transferase and glucose-6-phosphatase can be induced by OSM. It is also reported to cause accumulation of glycogen granules in developing hepatocytes and induce hepatic maturation by activating STAT3 [23]. Insulin activates PI3/AKT pathway which in turn has been reported to increase hepatic differentiation [24]. Figure 2.1 shows the differentiation of stem cells into different cell types and their validation by established markers of differentiation.

## 2.2 Stem Cells and Autophagy

Stem cells are unique in terms of their longevity since they persist throughout the life span of an organism. These cells are important for development, tissue renewal, and replacement, and dysfunctions in stem cell activity can lead to disease processes. Thus, a tight control over the process of cellular homeostasis is very important for maintenance of proper cell function. In this context, autophagy, which is essential for protein quality control mechanism, may play a crucial role [25].

Various experimental evidences have confirmed that self-renewal, pluripotency, differentiation, and quiescence in stem cells are dependent on the activation of autophagic process [26]. Self-renewal and differentiation both require a strict control of cellular remodeling based on protein turnover and lysosomal degradation of organelles. The elimination of defective macromolecules and organelles is a prerequisite to preserve the pluripotency during the long periods of quiescence often observed in stem cells. Thus, the basal autophagy in stem cells mediate clearance of damaged or defective intracellular proteins and organelles as well as maintains cellular remodeling and quality control through degradation of structural components [27]. Autophagic activity is also observed to increase in cells exposed to low oxygen or nutrient supply. Under normal physiological conditions, it is rare to observe autophagosomes in somatic cells [28]. Autophagy involves internalization of toxic proteins or organelles into double membrane autophagosome by implicating various conjugation systems involving ATG5, ATG7, ATG10, LC3, etc. [29, 30]. The latter process involves the fusion of autophagosome with lysosome which results into the degradation of toxicants by lysosome enzymes. The free fatty acids and amino acids generated as the end products of degradation are then recycled by cell for use in various cellular processes [31]. As stem cells age or encounter a pathological condition, they need to maintain their homeostasis by removal of toxic proteins and organelles. Not surprisingly, there are a number of evidences to account for the active role of autophagy in the regulation of different types of stem cells like various adult tissue stem including hematopoietic stem cells (HSCs) cells as well as embryonic stem cells (ESCs) and cancer stem cells (CSCs). Recent studies on aging muscle stem cells (MSC) and hematopoietic stem cells (HSC) have documented that dysfunction of autophagy activity in stem cells occurs with age. The results of these studies have confirmed that correct functioning of autophagic machinery in cell is necessary for maintenance of the proper blood system and muscle development. It also helps the survival of adult stem cells during metabolic stress [32].

Ma et al. demonstrated that autophagy is significantly reduced in aged mesenchymal stem cells derived from bone marrow (BMMSCs) compared with young BMMSCs. This may be associated with degenerative changes like loss of proper differentiation and reduced proliferation observed during aging. All these further contribute to age-related bone loss [33]. Another study demonstrated the decisive role of autophagy during the switch between the quiescence and senescence of muscle derived stem cells [32]. It was also proposed that reactivation of autophagy in the whole musculature could alleviate age-associated myofiber degeneration and mitochondrial dysfunction [34].

## 2.3 Autophagy in Stem Cell Lineage Commitment and Differentiation

At this stage, it is clear that impairment of autophagy in either physiologically aged cells or young cells could cause senescence, increased mitochondrial dysfunction, and oxidative stress, leading to a decline in cellular function. However, reestablishment of autophagy can reverse senescence and restore the regenerative functions in geriatric cells [32]. Stem cells are observed to be more resistant to cytotoxic drugs compared to differentiated cells. One of the reasons may be the high basal autophagy which helps these cells to overcome the therapy stress. However, when autophagy is blocked or differentiation is induced, the cells become sensitive [28]. In differentiated cells, autophagy is downregulated due to lower levels of ATG5 and Beclin1, two major regulators of autophagy. Various other autophagy associated proteins are involved in maintenance of stem cell state in different types of systems; for example, 200-kDa FAK-family interacting protein (FIP200) was reported to be crucial for maintenance of cell autonomous functions of fetal HSCs since its deletion caused severe anemia and perinatal lethality [35]. Another autophagy associated protein Atg7 was reported to be crucial for adult HSC maintenance in mouse [36]. Generally the stem/progenitor cells maintain their stemness by maintaining low levels of reactive oxygen species (ROS) due to decreased mitochondria and hence low dependence on oxidative phosphorylation, thus reducing the generation of ROS. Low levels of ROS help to prevent DNA damage in stem cells and in turn maintaining their stemness. There are evidences available where defective levels of Beclin and Atg resulted into increased tumorigenesis in mice models [37]. This was attributed to the increased defective mitochondria due to defective autophagy which resulted into increased ROS production inducing more DNA damage and ultimately tumor formation [38].

Basal autophagy is very low under normal conditions and it is activated in response to stress and various other extracellular cues. However, undifferentiated stem cells have high basal levels of autophagy which gets reduced to negligible levels as a stem cell differentiates terminally into various cellular derivatives [39]. Inhibition of autophagy in the adult stem cells may block differentiation. It was reported by Lee et al. that hypoxia-induced autophagy promoted maintenance and self-renewal of MSC [40]. Zhang et al. have also shown that activation of autophagy antagonized MSC apoptosis during hypoxia/serum deprivation, while inhibition of autophagy promoted the same [41]. Oliver et al. have provided a nice demonstration of autophagy downregulation in mesenchymal stem cells after osteoblastic differentiation [42]. In another study, the autophagic vacuoles were found to be in higher in dermal, epidermal, and hematopoietic stem cells (HSCs) as compared to acute promyelocytic leukemia (APL) cell line NB4 [28]. It was also observed that the lower level of autophagy in cells post differentiation was associated with decreased ATG5 and Beclin1 expression at transcription level. This study also reported the abolishment of self-renewal capacity of stem cells after downregulation of autophagy by pharmacological interventions. Autophagy is involved in alteration of the cell morphology and structure by remodeling during differentiation [43]. A defect in the autophagy proteins Atg5/Atg7 was shown to be associated with altered maturation of preadipocytes to adipocytes with accumulation of lipid droplets into small compartments instead of a single large droplet which is characteristic feature of normal adipocyte differentiation. This difference was found to be attributed to increased mitochondria number and hence enhanced β-oxidation which resulted into the depletion of fatty acids and hence impairment in triglyceride synthesis [44]. Apart from controlling the number of mitochondria, autophagy also controls differentiation by protein degradation and generation of amino acids. This was demonstrated beautifully in a study where Atg5 knockout resulted into abrogation of transition from four cells to eight cells stage in mouse embryo [45]. This was found to be correlated with decrease amino acid synthesis due to reduced

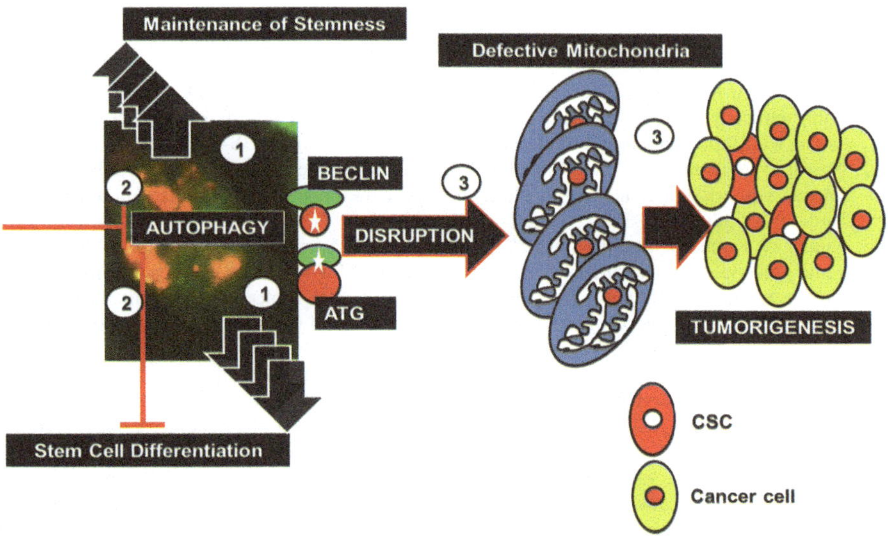


**Fig. 2.2** Autophagic tuning of Stemness: *1*. Undifferentiated stem cells generally maintain high levels of basal autophagy which gets reduced during stem cell differentiation, *2*. Inhibition of Autophagy can lead to compromised ability of a stem cell to undergo differentiation, *3*. Disruption in the autophagy related proteins can lead to defects in mitochondria which further leads to tumorigenesis with some cancer stem cells in interior which maintain the cancer pool

autophagy turnover and hence protein starvation. This was also found to be associated with accumulation of more maternal proteins in cytoplasm and hence defective zygote development. Thus, autophagy may function as a critical regulator of stem cell/progenitor cell function under both pathological and physiological conditions (Fig. 2.2).

In conclusion, we can say that autophagy is usually stimulated as a response to stress in differentiated cells, in adult stem cells, while elevated level of autophagic flux is a general phenomenon under physiological conditions. Based on the present evidences, autophagy is considered as a protective mechanism to ensure the well-being of adult stem cells. Autophagy ensures the restoration and subsequent survival of stem cells during stress like anticancer therapy. The survival of adult stem cells before they find their niche and engraft during transplantation procedures has been attributed to their relatively higher levels of basal autophagy. Thus, the recent advances in stem cell biology suggest that autophagy helps in fine-tuning the balance between quiescence, self-renewal, and differentiation of stem cells under normal physiological conditions as well as to counteract various stress stimuli.

The implication that autophagy is intimately involved in the maintenance of stem cell physiology adds a new level of cellular control. Autophagy regulation can provide a new operational basis for initiating strategies for efficient generation of induced pluripotent cells during somatic reprogramming. Finally, cells can overcome the barrier of cellular senescence by modulating the autophagic flux.

## References

1. Gandia C, Arminan A, Garcia-Verdugo JM, Lledo E, Ruiz A, Minana MD, et al. Human dental pulp stem cells improve left ventricular function, induce angiogenesis, and reduce infarct size in rats with acute myocardial infarction. Stem Cells. 2008;26(3):638–45.
2. Nosrat IV, Widenfalk J, Olson L, Nosrat CA. Dental pulp cells produce neurotrophic factors, interact with trigeminal neurons in vitro, and rescue motoneurons after spinal cord injury. Dev Biol. 2001;238(1):120–32.
3. Kumar A, Kumar V, Rattan V, Jha V, Bhattacharyya S. Secretome cues modulate the neurogenic potential of bone marrow and dental stem cells. Mol Neurobiol. 2017;54(6):4672–82.
4. Kumar A, Kumar V, Rattan V, Jha V, Pal A, Bhattacharyya S. Molecular spectrum of secretome regulates the relative hepatogenic potential of mesenchymal stem cells from bone marrow and dental tissue. Sci Rep. 2017;7(1):15015.
5. Bhattacharyya S, Kumar A, Lal Khanduja K. The voyage of stem cell toward terminal differentiation: a brief overview. Acta Biochim Biophys Sin (Shanghai). 2012;44(6):463–75.
6. Pera MF, Tam PP. Extrinsic regulation of pluripotent stem cells. Nature. 2010;465(7299):713–20.
7. Xiao Y, Peperzak V, van Rijn L, Borst J, de Bruijn JD. Dexamethasone treatment during the expansion phase maintains stemness of bone marrow mesenchymal stem cells. J Tissue Eng Regen Med. 2010;4(5):374–86.
8. Rosen ED, MacDougald OA. Adipocyte differentiation from the inside out. Nat Rev Mol Cell Biol. 2006;7(12):885–96.
9. Scott MA, Nguyen VT, Levi B, James AW. Current methods of adipogenic differentiation of mesenchymal stem cells. Stem Cells Dev. 2011;20(10):1793–804.
10. Ailhaud G. Adipose cell differentiation in culture. Mol Cell Biochem. 1982;49(1):17–31.
11. Qiu Z, Wei Y, Chen N, Jiang M, Wu J, Liao K. DNA synthesis and mitotic clonal expansion is not a required step for 3T3-L1 preadipocyte differentiation into adipocytes. J Biol Chem. 2001;276(15):11988–95.
12. Janderova L, McNeil M, Murrell AN, Mynatt RL, Smith SR. Human mesenchymal stem cells as an in vitro model for human adipogenesis. Obes Res. 2003;11(1):65–74.
13. Gurriaran-Rodriguez U, Al-Massadi O, Roca-Rivada A, Crujeiras AB, Gallego R, Pardo M, et al. Obestatin as a regulator of adipocyte metabolism and adipogenesis. J Cell Mol Med. 2011;15(9):1927–40.
14. Kim SP, Ha JM, Yun SJ, Kim EK, Chung SW, Hong KW, et al. Transcriptional activation of peroxisome proliferator-activated receptor-gamma requires activation of both protein kinase A and Akt during adipocyte differentiation. Biochem Biophys Res Commun. 2010;399(1):55–9.
15. Glaser T, Brustle O. Retinoic acid induction of ES-cell-derived neurons: the radial glia connection. Trends Neurosci. 2005;28(8):397–400.
16. Young JE, Martinez RA, La Spada AR. Nutrient deprivation induces neuronal autophagy and implicates reduced insulin signaling in neuroprotective autophagy activation. J Biol Chem. 2009;284(4):2363–73.
17. Wobus AM, Grosse R, Schoneich J. Specific effects of nerve growth factor on the differentiation pattern of mouse embryonic stem cells in vitro. Biomed Biochim Acta. 1988;47(12):965–73.
18. Janesick A, Wu SC, Blumberg B. Retinoic acid signaling and neuronal differentiation. Cell Mol Life Sci. 2015;72(8):1559–76.
19. Ying QL, Smith AG. Defined conditions for neural commitment and differentiation. Methods Enzymol. 2003;365:327–41.
20. Vollner F, Ernst W, Driemel O, Morsczeck C. A two-step strategy for neuronal differentiation in vitro of human dental follicle cells. Differentiation. 2009;77(5):433–41.
21. Snykers S, De Kock J, Tamara V, Rogiers V. Hepatic differentiation of mesenchymal stem cells: in vitro strategies. Methods Mol Biol. 2011;698:305–14.

22. Hu Z, Evarts RP, Fujio K, Marsden ER, Thorgeirsson SS. Expression of hepatocyte growth factor and c-met genes during hepatic differentiation and liver development in the rat. Am J Pathol. 1993;142(6):1823–30.
23. Kamiya A, Kinoshita T, Miyajima A. Oncostatin M and hepatocyte growth factor induce hepatic maturation via distinct signaling pathways. FEBS Lett. 2001;492(1-2):90–4.
24. Magner NL, Jung Y, Wu J, Nolta JA, Zern MA, Zhou P. Insulin and IGFs enhance hepatocyte differentiation from human embryonic stem cells via the PI3K/AKT pathway. Stem Cells. 2013;31(10):2095–103.
25. He C, Klionsky DJ. Regulation mechanisms and signaling pathways of autophagy. Annu Rev Genet. 2009;43:67–93.
26. Phadwal K, Watson AS, Simon AK. Tightrope act: autophagy in stem cell renewal, differentiation, proliferation, and aging. Cell Mol Life Sci. 2013;70(1):89–103.
27. Guan JL, Simon AK, Prescott M, Menendez JA, Liu F, Wang F, et al. Autophagy in stem cells. Autophagy. 2013;9(6):830–49.
28. Salemi S, Yousefi S, Constantinescu MA, Fey MF, Simon HU. Autophagy is required for self-renewal and differentiation of adult human stem cells. Cell Res. 2012;22(2):432–5.
29. Pan H, Cai N, Li M, Liu GH, Izpisua Belmonte JC. Autophagic control of cell 'stemness'. EMBO Mol Med. 2013;5(3):327–31.
30. Rubinsztein DC, Marino G, Kroemer G. Autophagy and aging. Cell. 2011;146(5):682–95.
31. Wirawan E, Vanden Berghe T, Lippens S, Agostinis P, Vandenabeele P. Autophagy: for better or for worse. Cell Res. 2012;22(1):43–61.
32. Garcia-Prat L, Martinez-Vicente M, Perdiguero E, Ortet L, Rodriguez-Ubreva J, Rebollo E, et al. Autophagy maintains stemness by preventing senescence. Nature. 2016;529(7584):37–42.
33. Ma Y, Qi M, An Y, Zhang L, Yang R, Doro DH, et al. Autophagy controls mesenchymal stem cell properties and senescence during bone aging. Aging Cell. 2017;17(1)
34. Carnio S, LoVerso F, Baraibar MA, Longa E, Khan MM, Maffei M, et al. Autophagy impairment in muscle induces neuromuscular junction degeneration and precocious aging. Cell Rep. 2014;8(5):1509–21.
35. Liu F, Lee JY, Wei H, Tanabe O, Engel JD, Morrison SJ, et al. FIP200 is required for the cell-autonomous maintenance of fetal hematopoietic stem cells. Blood. 2010;116(23):4806–14.
36. Mortensen M, Soilleux EJ, Djordjevic G, Tripp R, Lutteropp M, Sadighi-Akha E, et al. The autophagy protein Atg7 is essential for hematopoietic stem cell maintenance. J Exp Med. 2011;208(3):455–67.
37. Yue Z, Jin S, Yang C, Levine AJ, Heintz N. Beclin 1, an autophagy gene essential for early embryonic development, is a haploinsufficient tumor suppressor. Proc Natl Acad Sci U S A. 2003;100(25):15077–82.
38. Jin S. Autophagy, mitochondrial quality control, and oncogenesis. Autophagy. 2006;2(2):80–4.
39. Revuelta M, Matheu A. Autophagy in stem cell aging. Aging Cell. 2017;16(5):912–5.
40. Pyo JO, Nah J, Jung YK. Molecules and their functions in autophagy. Exp Mol Med. 2012;44(2):73–80.
41. Zhang Q, Yang YJ, Wang H, Dong QT, Wang TJ, Qian HY, et al. Autophagy activation: a novel mechanism of atorvastatin to protect mesenchymal stem cells from hypoxia and serum deprivation via AMP-activated protein kinase/mammalian target of rapamycin pathway. Stem Cells Dev. 2012;21(8):1321–32.
42. Oliver L, Hue E, Priault M, Vallette FM. Basal autophagy decreased during the differentiation of human adult mesenchymal stem cells. Stem Cells Dev. 2012;21(15):2779–88.
43. Vessoni AT, Muotri AR, Okamoto OK. Autophagy in stem cell maintenance and differentiation. Stem Cells Dev. 2012;21(4):513–20.
44. Singh R, Xiang Y, Wang Y, Baikati K, Cuervo AM, Luu YK, et al. Autophagy regulates adipose mass and differentiation in mice. J Clin Invest. 2009;119(11):3329–39.
45. Tsukamoto S, Kuma A, Mizushima N. The role of autophagy during the oocyte-to-embryo transition. Autophagy. 2008;4(8):1076–8.

# Chapter 3
# Role of Autophagy in Aging of Hematopoietic Stem Cells and Their Niche: Relevance in Clinical Transplantations and Regenerative Medicine

**Rohan S. Kulkarni, Manmohan Bajaj, and Vaijayanti P. Kale**

**Abstract** Stem cell transplantation (SCT) is the only curative therapy for various malignant as well as nonmalignant disorders like leukemia, lymphoma, and aplastic anemia. Efficacy of clinical transplantation critically depends on functionality of the hematopoietic stem cells (HSCs) present in the donor graft, and therefore, it is essential that every precaution should be taken to ensure that the donor HSCs possess good engraftment ability. Aging is known to cause decrease in the functionality of HSCs, and, therefore, clinicians prefer younger donors. However, if only a single HLA-matched aged donor is available, the patient may not be able to avail the life-saving SCT treatment. It is therefore necessary to device strategies to reverse the aging of HSCs. Here we shall review the available literature on the role of autophagy in the functionality of HSCs and their niche cells. We shall also discuss the importance of induction of autophagy in reversal of aging-mediated dysfunction in them and its relevance in clinical SCT and other regenerative medicine protocols.

## 3.1 Introduction

Stem cell transplantation (SCT) is the only curative therapy for various malignant and nonmalignant diseases and it has been successfully applied for several decades in the treatment of these diseases. However, the efficacy of SCT critically depends on the quality and quantity of the hematopoietic stem cells (HSCs) present in the

R. S. Kulkarni · M. Bajaj
Stem Cell Lab, National Centre for Cell Science, Pune, India

V. P. Kale (✉)
Stem Cell Lab, National Centre for Cell Science, Pune, India

Symbiosis Centre for Stem Cell Research, Symbiosis School of Biological Sciences, Symbiosis International (Deemed University), Pune, India
e-mail: vpkale@nccs.res.in; vaijayanti.kale@ssbs.edu.in



K. Turksen (ed.), *Autophagy in Health and Disease*, Stem Cell Biology and Regenerative Medicine, https://doi.org/10.1007/978-3-319-98146-8_3

graft. Aging is known to cause a significant decrease in the functionality of HSCs, and therefore, clinicians prefer younger donors. However, if only single HLA-matched aged donor is available, the patient may not be able to avail this life-saving treatment.

The aging-mediated decrease in the functionality of HSCs has been correlated with a decrease in the levels of autophagy in them [1], suggesting that a strategy that leads to an increase in the autophagy levels in the aged HSCs, and also in their niche cells, may prove to be useful in reversing the aging-mediated loss of HSC functionality, and thereby increase the donor cohort. We have shown that microvesicles (MVs) isolated from young mesenchymal stromal cells (MSCs) harbor autophagy-inducing mRNAs, and treatment of aged HSCs with these "young" MVs boosts their engraftment capacity and also reduces their myeloid bias [2]. We have also shown that it is possible to rejuvenate the aged mesenchymal stromal cells (MSCs) by pharmacologically inhibiting the AKT signaling in them.

In the present chapter we review the literature on the role of autophagy in the functionality of HSCs. We also discuss the relevance of reversal of autophagy in aged HSCs in clinical transplantations and regenerative medicine.

## 3.2 Autophagy

Autophagy is a major proteolytic system implicated in proteostasis and protein quality control [3]. It involves a complex network of proteins working in multiple sub-complex systems in a linear fashion. The word autophagy is derived from Greek: auto, meaning "self" and phagy meaning "to eat." In this process, cells' own components degrade through the lysosomal enzymatic pathway [4]. It is characterized by the engulfment of the targeted components in autophagosomes followed by their fusion with lysosomes, leading to degradation of the contents by resident hydrolases. The produced catabolites are rapidly made available in the cytoplasm for recycling [5]. Autophagy is an evolutionary conserved process, and targets and degrades misfolded proteins or functionally impaired organelles, thus preventing the toxic effects due to their accumulation [3]. In somatic cells, the quality control of long-lived proteins and organelles is ensured by autophagy. Thus, autophagy is predominantly a cytoprotective process, rather than self-destructive process [6].

In fact, the process of autophagy is majorly divided into three types, viz., chaperone-mediated autophagy, microautophagy, and macroautophagy. In chaperone-mediated autophagy, small protein molecules are recognized based upon specific signaling peptide that serves as a signal for degradation of these molecules. Specific type of proteins such as heat shock proteins (HSPs) and chaperones bind to the protein and translocate it to the lysosomal degradation site. Small cytoplasmic molecules and organelles are directly engulfed by lysosomal membrane for degradation for the process of microautophagy, whereas in the case of macroautophagy larger cell organelles are enclosed in phagophore to form an autophagosome, which then fuses with the large lysosome for degradation of these components [7]. Out of these three types, macroautophagy is considered to be the major type. Many autophagy-related (Atg) proteins are known to play an important role in autophagosome formation. Atg1–18, except Atg11, 15, and 17, are considered to be the core Atg proteins. These proteins form autophagosome with the help of other Atg proteins. Autophagy initiation requires Vsp34 complex made up of Atg 6, Atg 14, and Vsp15. Accessory proteins such as Atg 5, 12, 16 and FIP200 along with Atg 1 form the Atg1 complex. This marks the initiation stage of the autophagosome formation and elongation of this complex interaction of Atg12 with Atg 5, 7, 10, and 16L1. On completion of the process, the Atg5–Atg12 conjugate dissociates from the phagosome complex. In the next step, Atg8 (LC3) interacts with phosphatidylethanolamine (PE) to form LC3-II. These LC3-bearing autophagosomes are then transported to the lysosomes for degradation of macromolecules [8]. This process is activated in response to extracellular or intracellular stress and signals such as starvation, growth factor deprivation, ER stress, and pathogen infection, and thus serves as an important adaptive response pathway against stress.

Malfunctioning of autophagy contributes to a variety of diseases including cancer [9], neurodegeneration [10], and cardiovascular disorders [11]. Autophagy is required for cellular homeostasis and various cellular processes such as self-renewal, differentiation, and cell survival [12]. Focal adhesion kinase protein of 200 kDa (FIP200) [13] and autophagy proteins encoded by Atg7 [14] have been shown to be essential for maintenance of HSC functionality by induction of autophagy. Activity of Sirtuin family of protein, Sirt1 is reported to influence the autophagy directly by acting on the components of the autophagy machinery, which was confirmed by coimmunoprecipitation of Sirt1 with Atg5, Atg7, and LC3 [15]. Sirt1 can also enhance the expression of the components of the autophagic machinery genes by activating them through Foxo transcription factors [16]. The role of Foxo3 and Foxo1 in promoting autophagy and cardiomyocyte survival under oxidative stress condition has also been well demonstrated [17, 18]. Foxo1 can regulate multiple autophagy-related genes such as Beclin1, Lc3, Atg12, Bnip3, and Ulk2 [19]. Similarly, Sirt2 can mediate the regulation of autophagy-related gene Atg7 through Foxo1 acetylation.

Autophagy processes are known to maintain the cellular homeostasis by maintaining the intercellular quality control [7]. Alteration in autophagy process is seen not only to modulate the cellular fates, but also the organismal fates as a whole. Activity of autophagy-related and autophagy-inducing proteins reduces during aging. On the other hand, increase in autophagy by pharmacological means extends the organismal life [3]. Loss of proteostasis and mitochondrial dysfunction are considered as two major hallmarks of aging [20]. Autophagy is required for maintaining the protein quality and mitochondrial functionality of the cell to combat aging-related dysfunctions. Many longevity-promoting molecules like rapamycin and resveratrol are known to stimulate this process by increasing the autophagy levels [21]. In case of HSCs, autophagy has been identified as an essential mechanism for protection against metabolic stress and cytokine withdrawal. Caloric restriction is seen to robustly induce Foxo3a-directed protective autophagy program in them [22]. Several studies implicate autophagy-inducing molecules for overcoming the ageing-related anomalies in various types of cells, including stem cells to extend the applications in anti-aging and regenerative therapies. Thus, autophagy and its regulation are important aspects in maintaining the functionality and the quality of HSCs during aging.

## 3.3 Autophagy in HSC Functions

HSCs are perhaps the oldest known and the most well characterized stem cells. Bone marrow or stem cell transplantation (BMT/SCT) is the only curative therapy for several malignant and nonmalignant diseases. As against BMT, wherein bone marrow harvested from the iliac crest is used as a graft, in SCT, granulocyte colony-stimulating factor (G-CSF)-induced mobilized peripheral blood is used for transplantation. Transplantation is a stressful condition for the infused HSCs, as the recipients are subjected to severe myeloablative conditioning. In contrast to differentiated cells, where autophagy is usually induced as a consequence of stress, high autophagic activity is a general phenomenon of adult HSCs under physiological conditions. It may represent an immediately available safety mechanism to ensure cell repair and subsequent survival under stress conditions, including anti-cancer therapy. Interestingly, G-CSF-induced mobilization of HSCs has been shown to activate autophagy process in them, which protects them from stress-induced apoptosis [23].

Stem cells are long-lived and persist throughout the adult life of an organism, and therefore, the quality control mechanisms and maintenance of cellular homeostasis would be crucial for the maintenance of these cells. Thus, autophagy is expected to play an important role in the normal function of stem cells and associated diseases. Indeed available data suggest that the unique properties of stem cells (self-renewal, pluripotency, differentiation, and quiescence) are dependent on the activation of the autophagy process [24–26]. FIP200 (200 kDa focal adhesion kinase family interacting protein) was shown to be essential not only for the induction of autophagy, but also for the maintenance and the function of HSCs in vivo [13]. Importance of autophagy in protecting the HSCs from metabolic stress induced by in vitro cytokine withdrawal or in vivo calorie restriction has been demonstrated [22]. The transcription factor FOXO3A was found to be essential for maintaining a pro-autophagy gene program that poises HSCs for rapid autophagy induction. Interestingly, only the HSCs, but not their myeloid progeny, could mount a robust autophagy response in response to the stress. HSCs in humans have been shown to have constitutively higher levels of autophagy, which plays a role in differentiation and resistance to cytotoxic drugs in these cells [12]. Autophagy plays an important role in blood cell development as evidenced by the fact that mice lacking essential autophagy genes like Atg7 showed deregulated fetal and neonatal hematopoiesis. Mice deficient in Atg7 developed severe anemia due to accumulation of damaged mitochondria and consequent cells death. Similarly, it was also demonstrated that the absence of Atg7 in the hematopoietic system leads to defective removal of mitochondria in erythroid cells, due to which the mice develop severe progressive anemia [14]. A hematopoietic chimeric mouse generated with Atg5$^{-/-}$ fetal liver resulted in T lymphopenia, suggesting a crucial role of autophagy in T cell survival and function. These mice also showed reduced numbers of thymocytes and B lymphocytes, suggesting Atg5's importance in homeostatic proliferation of lymphocytes [27]. Furthermore, Atg7$^{-/-}$ mice fail to form secondary colonies during colony-forming cell (CFC) assays, suggesting a defect in self-renewal [14]. Likewise, a pharmacological inhibition of autophagy in human HSCs leads to their failure to form colonies in colony forming unit assay [25]. Atg7 deficiency also leads to severe lymphopenia as a result of mitochondrial damage followed by apoptosis in mature T lymphocytes. Interestingly, myeloid (CD11b$^+$) and dendritic cells (CD11c$^+$ DCs) were not affected due to ATG7 deficiency. Autophagy-mediated mitochondrial clearance (i.e., mitophagy) may affect HSC mitochondrial content, thereby influencing the fate of HSCs and maintenance of hematopoietic homeostasis [28]. Metabolic state of HSCs is known to play an important role in maintaining stemness of the HSCs [29]. It has been shown that fetal HSCs have high mitochondria and OxPhos in them and these HSCs switch to glycolysis at around 4 weeks postnatal period [30, 31]. Further studies done with HSCs isolated based on their mitochondrial membrane potential showed that LT-HSCs with low mitochondrial potential have higher repopulation ability, as compared to LT-HSCs having high mitochondrial membrane potential. It was further demonstrated that HSCs with low mitochondrial mass have high levels of autophagy protein LC3B, further establishing the role for autophagy in modulating metabolism of HSCs and their repopulation ability [32]. These reports underscore the importance of autophagy in HSC maintenance, self-renewal and functionality.

## 3.4 Aging of HSCs

Aging takes toll in all cell types including stem cells. HSCs, which are the main source of blood-forming cells, are known to accumulate age-related changes in their functionality. Analysis of peripheral blood cells of the aged organisms clearly showed an increased frequency of myeloid cells and decreased frequency of lymphoid cells [33, 34]. The aged HSCs are known to give rise to increased number of cobblestone area forming colonies (CAFCs) in in-vitro functionality assay [35]; however, the in vivo functionality of the HSCs determined by its capacity to reconstitute the blood system in the recipient organism is found to be diminished. The transplantation experiments performed by various groups of researchers have confirmed the compromised ability of aged HSCs to completely reconstitute the blood system in the recipients. On these lines, the long-term engraftment potential of aged HSCs was seen to be drastically diminished as determined by transplantation of the donor HSCs from primary recipients to secondary recipients [34, 36].

The aged BM shows accumulation of dysfunctional HSCs and this might be the outcome of reduction in cell clearance capacity of aged HSCs. Increase in senescence and reduced apoptosis are the main drivers of the characteristic aged HSC accumulation [37, 38]. The expression of apoptosis-related genes, *Fas* and *Bad* are reported to be decreased in the aged HSCs [38]. Quiescent HSCs accumulate DNA damage due to attenuation of DNA damage response during aging [34]. Along with this loss of proteostasis, increased mitochondrial dysfunction is considered to be the major driver of HSC aging [20].

Autophagy is responsible for maintaining the functional HSCs in low metabolic stress conditions as compared to the short lived populations such as myeloid progenitors. Authors found that FOXO3A—an important regulator of autophagy—is crucial for regulating the pro-autophagic gene expression and is required to maintain the stemness of HSCs. Aged HSCs were initially shown to retain this pro-autophagic program in them [22]. But the further investigation of the aged HSC compartment showed that the autophagy program is reduced in most of the aged HSCs, but about one third of this population has a higher active autophagy levels and is responsible for cumulative functioning of the aged HSC compartment [1]. Further, they showed that the aged HSCs with active autophagy are more functional, as compared to those having low autophagy levels. This was achieved by actively suppressing the metabolic stress by eliminating the healthy mitochondria resulting in increased quiescence of HSCs. The HSCs which are maintained in the functional state ex vivo in the presence of mTORC1 and GSK-3 inhibitor can be identified by their active autophagy state [39]. We have also shown that the aged HSCs have decreased pro-apoptotic gene signature, which can be restored upon exposing them to young microenvironment, thereby increasing the functionality of these HSCs [2]. Moreover, a conditional deletion of ATG7 in murine HSCs makes them dysfunctional, leading to death of these mice within weeks after the deletion. These mice also show severe myeloproliferative characters, like those seen in aged HSCs [14]. Several cell- intrinsic mechanisms, including reduced autophagy, lead to the aging of HSCs [40]. Reduced autophagy has been associated with accelerated aging, whereas stimulation of autophagy might have potent anti-aging effects [21]. Lack of autophagy in hematopoietic system has been shown to lead to loss of HSC function and promote myeloid cell proliferation [14], a situation akin to that seen in aging individuals. It has been demonstrated that high levels of reactive oxygen species, generated by mitochondria, accumulate in aged HSCs and compromise their functioning [41, 42]. HSCs from $Atg^{-/-}$ mice show high levels of ROS suggesting a role of autophagy in aging of HSCs [14]. Recently, Ho et al. [1] showed that loss of autophagy in HSCs causes accumulation of mitochondria and an activated metabolic state, which drives accelerated myeloid differentiation mainly through epigenetic deregulations, and impairs their self-renewal activity and regenerative potential. They further showed that approximately one-third of aged hematopoietic stem cells exhibit high autophagy levels and maintain a low metabolic state with robust long-term regeneration potential similar to healthy young HSCs. These results demonstrate an important role of autophagy in preserving the regenerative capacity of aged HSCs. Likewise, a subset of aged HSCs having higher autophagy levels was shown to engraft efficiently, as compared to their counterparts having low levels of autophagy [1], underscoring the importance of autophagy in engraftment ability of HSCs.

In our recent study, we have shown that the microvesicles (MVs) secreted by young MSCs are enriched with mRNAs coding for autophagy, and when aged HSCs are treated with these young MVs they show significantly improved engraftment capacity and reduced myeloid bias [2]. Collectively, these studies emphasize the role of autophagy in the engraftment ability of HSCs.

Reducing mitochondrial stress in aged HSCs can reverse their loss of functionality [42]. Mitochondrial dysfunction causes multiple hematopoietic defects that are typically seen in the elderly, but HSCs themselves appear to be relatively resistant [43]. Taken together, cellular metabolism, controlled by mitochondrial status and reactive oxygen species and mTORsignaling, plays an important role in maintaining HSC function throughout life, but the molecular cause of age-dependent metabolic derailment remains unclear. Importantly, however, pharmacological interventions in these signaling pathways are feasible and may be exploited to restore function in aged HSCs.

## 3.5 Aging of MSCs

MSCs and MSC-derived differentiated progenies like osteoblasts and adipocytes are the main components of HSC niche and any change in their composition and functionality in the niche alters the HSC functionality during aging. In the study by Stolzing et al. [44], the number of fibroblast colony-forming units (CFU-F), which are the direct indicators of frequency and proliferative potential of human MSCs decreased with age, indicating reduced frequency and loss of proliferative potential of aged MSCs. Further, apoptosis was found to be increased significantly along with increasing ROS levels which are correlated with decreased SOD expression in aged MSCs. An evident increase of cellular stress induction in aged MSCs [44], genomic DNA aberrations assessed by γ-H2AX staining and reduction in telomere length were reported as the characteristic features of aged MSCs. Expression of Connexin-43, the gap junction protein having a crucial role in maintaining lower levels of cellular ROS [45], was found to be low in the aged MSCs [46]. The aged MSCs are reported to be skewed toward adipogenesis at the expense of osteogenic or chondrogenic differentiation [47, 48]. As adipocytes have been proven to be negative regulators of HSC functionality [49] and also of B-lymphopoiesis [50], the change in the niche from osteoblast-dominant to adipocyte-dominant form during aging might be the key factor for decrease in HSC functionality during aging. Taken together, the HSC niche is reported to undergo drastic changes during aging, and thus, loses its ability to support the maintenance, self-renewal, and functionality of HSCs. The compositional and functional properties of the niche change during aging. These changes are thus thought to be reflected in HSC functionality and further deteriorate the HSCs qualitatively. Decreased formation of bones, increased adipogenesis, changes in extracellular matrix (ECM) components, and secretion of HSC-supportive cytokines are some of the important changes taking place in bone marrow during aging [51–53].

Recently, Pennock et al. [54] showed that the specific 3D growth environment promoting autonomous autophagy flux rejuvenated aged human MSCs through cytoplasmic remodelling, mitochondrial regression and shifting of the metabolic flux from oxidative phosphorylation to anaerobic one [54]. Autophagy was shown to be reduced in aged MSCs. Another report suggests that the age- related decline in cellular functions of MSCs corresponds to the decline in their differentiation potentials, which was attributed to the poor cytoskeletal dynamics and elevated exposure to ROS [55]. Autophagy is one of the major cellular processes that regulates MSC functionality and their differentiation. MSCs have been known to exhibit high level of constitutive autophagy, and its suppression affects survival and differentiation of human MSCs [56]. According to another report, activation of autophagy is essential for neural differentiation of mesenchymal stem cells [57]. Autophagy is known to protect the MSCs in stress conditions [58]. Induction of autophagy is also an important initial process for commitment and differentiation of MSCs into the osteogenic lineage [59]. Autophagy was shown to be reduced in aged MSCs, in turn reducing their proliferative potential and shifting their differentiation ability from osteogenic to adipogenic lineage [60]. It was also clear from this study that inhibiting autophagy by 3-methyladenine (3MA) induced the aged-like characteristics in young MSCs; on the other hand activation of autophagy by the m-TOR inhibitor rapamycin rejuvenated the aged MSCs by altering reactive oxygen species (ROS) and p53 levels.

Hypoxia-induced autophagy pathway has been shown to promote MSC proliferation [61]. Other molecules influencing MSCs' aging include stromal cell derived factor-1β (SDF1β), which also supports cell survival by enhancing autophagy process in them [62]. Induction of autophagy by serum starvation or by rapamycin treatment was studied and was shown to prevent MSCs from the irradiation injury and assist in the maintenance of their stemness by decreasing the DNA damage caused through intracellular ROS levels [63]. In a study regarding $CCl_4$-induced hepatic damage model, transplantation of MSCs led to elevation of autophagy levels for hepatic regeneration through systemic pathways involving HIF-1α [64]. Moreover, attenuation of AKT/mTOR pathways, known to inhibit autophagy, showed attenuation of age-related changes and enhanced proliferative capacity, clonogenic frequency, and osteogenic potential in MSCs [65]. Alternative studies on hypoxia pretreatment showed increased MSC survival, which promoted angiogenesis by enhancing autophagy through elevated expression of HIF-1α in them [66]. MSCs transplanted in IRI lung injury lead to activation of the autophagy pathway. Overall the therapeutic potential of BM-MSCs in various damage models is reported to increase by activation of autophagy in them [61, 66, 67].

MSCs form an important constituent of the bone marrow microenvironment and are known to form a special HSC niche [68]. Level of autophagy in mesenchymal stem/stromal cells (MSCs) is reported to be high and it becomes undetectable when these cells are differentiated into osteoblasts [56]. Hypoxia is known to enhance HSC-supportive nature of MSCs [61] and this could be related to hypoxia-mediated increase in the autophagy levels in them. Activation of autophagy antagonized, while inhibition of autophagy promoted apoptosis of MSCs during hypoxia/serum deprivation [58], indicating that autophagy is important in survival of MSCs under stress conditions.

Autophagy is susceptible to mTOR activation pathway, a downstream event of PI3K/AKT pathway [69]. Activated mTOR directly inactivates critical proteins involved in the activation of autophagy. In our previous work, we have shown that bone marrow-derived mesenchymal stromal cells (MSCs) gain activated AKT signaling as a consequence of aging, which leads to a reduction of autophagy-related genes in them, and the partitioning of these mRNAs into their microvesicles also goes down. Additionally, partitioning of miRNAs like miR-17 and 32b, which are negative regulators of autophagy-related mRNAs, into their exosomes goes up. This leads to aging of HSCs, as they not only receive the MVs which are deficient in autophagy-inducing mRNAs, but additionally, also receive exosomes containing miRNAs that degrade autophagy-inducing mRNAs [2]. This niche-mediated aging of HSCs involving autophagy process underscores the importance of autophagy in HSC functionality.

## 3.6 Relevance in Clinical Transplantation

The focus of the stem cell biology research has been development of techniques for rejuvenation of aged HSCs using different pharmacological or cellular molecules involved in modulation of HSC functioning, intrinsically or extrinsically. The first hope of rejuvenation of HSCs came from the interesting discovery of induced pluripotent stem cells (iPSCs), where the introduction of four transcription factor genes in the differentiated somatic cells de-differentiated them into stem cells [70]. The results suggested that the epigenetic modulation of the aged cells could result in increasing their functionality and induce their rejuvenation. The reprogramming of aged HSCs into iPSCs and their re-differentiation into HSCs, showed rejuvenating effect with increase in their repopulation ability, suggesting that the epigenetic component of HSCs is amenable to be reprogrammed to young-like state [71]. Satb1 is yet another molecule, which regulates chromatin organization and regulates lymphoid progenitor cells. Expression of Satb1 was found to be decreased in aged cells, while over-expression of Satb1 in aged HSCs resulted in increased lymphoid progeny in vitro [72], thus arising as a target for studies related to HSC rejuvenation. On the contrary, Per2 was seen to be increased in the aged lymphoid-biased HSCs and resulted in activation of apoptosis mediated through DDR and p53 knockout mice. Knocking down Per2 gene in aged HSCs attenuated their lineage skewing and maintained telomere length [73]. Overexpression of Sirt3, the mitochondrial member of sirtuin family, in aged HSCs decreases ROS production in HSCs and increases their reconstitution capacity [74]. The upregulation of Sirt7 in aged HSCs reduces mitochondrial protein folding stress, reduces myeloid skewing, and improves regenerative potential; the same effect is observed when NRF1 is downregulated in aged HSCs, confirming that the Sirt7-NRF-mediated interplay plays a crucial role in rejuvenation of aged HSCs [42]. mTOR pathway has been implicated in amplifying the aging effect in various tissues and its inhibition by rapamycin or mTOR inhibitors reduces aging effects and promotes longevity [75]. The treatment of aged mice with rapamycin for 6 weeks reduces *p16ink4a* and *p19Arf* expression in HSCs, reduces HSC frequency in bone marrow and increases in vivo regenerative capacity of aged HSCs [76]. Prolonged fasting for 72 h of aged mice has also been reported to reduce circulating IGF-1 levels and PKA activity, which led to signal transduction changes in their HSCs resulting in increased stress resistance, self-renewal and decreased myeloid progeny formation from them [77]. A long-term calorie restriction with about 75% of the normal diet post-pones senescence in hematopoietic compartment, and reduces the aging-related changes in the peripheral blood compartment, increases the function HSC frequency and improves repopulation capacity of aged HSCs from 25 months aged Balb/c mice. Short-term calorie restriction (for 5 months) shows the similar effect on aging HSC compartment in aged mice [78]. The aged HSCs are also reported to have increased Cdc42 activity resulting in loss of their polarity. This increase in the Cdc42 activity is mainly due to the shift from canonical to noncanonical Wnt signaling that shifts from Wnt3a to Wnt5a during aging [79]. Treatment with pharmacological inhibitor of Cdc42, CASIN increases the polarization of aged HSCs with respect to Cdc42, tubulin and Per-2. CASIN treatment also increases the self-renewal capacity of HSCs, decreases the myeloid cell output in transplanted recipients and rejuvenates HSCs [80]. As senescent cells also affect the other healthy cells through altered cytokine signaling and secretion of senescence-associated bodies, the depletion of senescent cells by treatment of anti-apoptotic protein inhibitor ABT263 increases the HSC functionality and reduces myeloid lineage skewing of aged HSCs [81]. Most importantly, in our recent report we showed that the aged HSCs can be successfully rejuvenated by co-culturing them with young MSCs. Furthermore, we also showed that rescued aged MSCs can be used to rejuvenate aged HSCs after inactivation of PI3K/AKT signaling in them [2].

## 3.7 Perspective

As mentioned before, autophagy plays an important role in the functionality of HSCs, and therefore, induction of autophagy in HSCs could form a very effective way to improve the efficacy of transplantation. In addition to several factors that modulate autophagy levels in the HSCs, the stromal cells appear to play a crucial role in this process. HLA-matched donors are becoming harder to find in this modern era of nuclear families. In this scenario, elimination of older donors due to their aging-related loss of HSC functionality further limits the donor cohort. Our work has clearly demonstrated that induction of autophagy by treating them with microvesicles collected from young MSCs could rejuvenate the aged HSCs and increase the donor cohort. Thus, in vitro rejuvenation of HSCs having compromised autophagy process either due to aging or due to exposure to chemotherapeutic drugs, by exposing them to MSC-derived microvesicles enriched in autophagy-inducing mRNAs appears to be a simple and effective approach to improve outcome of clinical transplants.

MSCs are being used in several regenerative therapies, and they have been the subject of several clinical trials, but the outcomes of advanced clinical trials have fallen short of expectations raised by encouraging preclinical animal data in a wide array of disease models [82]. The dissonance between mouse and human clinical outcomes has been thought to be due to the apparent discrepancies of immune compatibility, dosing, and fitness of culture-adapted MSCs. Typically, MSCs are sourced from varied sources like bone marrow, adipose tissue, dental pulp, placenta, and cord; the choice primarily appears to be dictated by the ease of obtaining the tissue. The culture expanded MSCs are usually characterized by a panel of phenotypic markers and their tri-lineage differentiation potential, but there are no studies that deal with identification of signaling pathways prevailing in them. Also, there has been no significant attempt made to characterize the macromolecular profile of the extracellular vesicles secreted by them with respect to their culture condition, source or their biochemical makeup. Our work underscores the importance of these aspects in characterizing the MSCs for clinical applications [2, 83]. Our data clearly show the need to characterize MSCs with respect to their signaling mechanisms prevailing in them, and macromolecular content of their EVs needs to be analysed. It is possible to develop specific culture conditions to get desired macromolecular profile of their EVs. We further propose that use of EVs, rather than MSCs, could also solve the problem of dosing. Being small in size, these vesicles might escape getting trapped in the filter organs and reach in sufficient quantities to the target organs. Also, these EVs could be cryopreserved and used clinically to treat the HSCs in vitro prior to transplantation. For other applications, it may also be possible to inject the EVs in damaged organs to improve their functionality. Such approach could enhance the outcome of regenerative therapies using MSCs.

## References

1. Ho TT, Warr MR, Adelman ER, Lansinger OM, Flach J, Verovskaya EV, Figueroa ME, Passegué E. Autophagy maintains the metabolism and function of young and old stem cells. Nature. 2017;543(7644):205.
2. Kulkarni R, Bajaj M, Ghode S, Jalnapurkar S, Limaye L, Kale VP. Intercellular transfer of microvesicles from young mesenchymal stromal cells rejuvenates aged murine hematopoietic stem cells. Stem Cells. 2018;36(3):420–33.
3. Rubinsztein DC, Mariño G, Kroemer G. Autophagy and aging. Cell. 2011;146(5):682–95.
4. Levine B, Yuan J. Autophagy in cell death: an innocent convict? J Clin Invest. 2005;115(10):2679–88.
5. Wirawan E, Berghe TV, Lippens S, Agostinis P, Vandenabeele P. Autophagy: for better or for worse. Cell Res. 2012;22(1):43.
6. Kroemer G, Levine B. Autophagic cell death: the story of a misnomer. Nat Rev Mol Cell Biol. 2008;9(12):1004.
7. Mizushima N, Komatsu M. Autophagy: renovation of cells and tissues. Cell. 2011;147(4):728–41.

8. Ravikumar B, Sarkar S, Davies JE, Futter M, Garcia-Arencibia M, Green-Thompson ZW, Jimenez-Sanchez M, Korolchuk VI, Lichtenberg M, Luo S, Massey DC. Regulation of mammalian autophagy in physiology and pathophysiology. Physiol Rev. 2010;90(4):1383–435.
9. Choi KS. Autophagy and cancer. Exp Mol Med. 2012;44(2):109.
10. Nah J, Yuan J, Jung YK. Autophagy in neurodegenerative diseases: from mechanism to therapeutic approach. Mol Cells. 2015;38(5):381.
11. Mei Y, Thompson MD, Cohen RA, Tong X. Autophagy and oxidative stress in cardiovascular diseases. Biochim Biophys Acta. 2015;1852(2):243–51.
12. Salemi S, Yousefi S, Constantinescu MA, Fey MF, Simon HU. Autophagy is required for self-renewal and differentiation of adult human stem cells. Cell Res. 2012;22(2):432.
13. Liu F, Lee JY, Wei H, Tanabe O, Engel JD, Morrison SJ, Guan JL. FIP200 is required for the cell-autonomous maintenance of fetal hematopoietic stem cells. Blood. 2010;116(23):4806–14.
14. Mortensen M, Soilleux EJ, Djordjevic G, Tripp R, Lutteropp M, Sadighi-Akha E, Stranks AJ, Glanville J, Knight S, Jacobsen SE, Kranc KR. The autophagy protein Atg7 is essential for hematopoietic stem cell maintenance. J Exp Med. 2011;208:455.
15. Lee IH, Cao L, Mostoslavsky R, Lombard DB, Liu J, Bruns NE, Tsokos M, Alt FW, Finkel T. A role for the NAD-dependent deacetylase Sirt1 in the regulation of autophagy. Proc Natl Acad Sci. 2008;105(9):3374–9.
16. Daitoku H, Sakamaki JI, Fukamizu A. Regulation of FoxO transcription factors by acetylation and protein–protein interactions. Biochim Biophys Acta. 2011;1813(11):1954–60.
17. Sengupta A, Molkentin JD, Paik JH, DePinho RA, Yutzey KE. FoxO transcription factors promote cardiomyocyte survival upon induction of oxidative stress. J Biol Chem. 2011;286(9):7468–78.
18. Sengupta A, Molkentin JD, Yutzey KE. FoxO transcription factors promote autophagy in cardiomyocytes. J Biol Chem. 2009;284(41):28319–31.
19. Kroemer G, Mariño G, Levine B. Autophagy and the integrated stress response. Mol Cell. 2010;40(2):280–93.
20. López-Otín C, Blasco MA, Partridge L, Serrano M, Kroemer G. The hallmarks of aging. Cell. 2013;153(6):1194–217.
21. Madeo F, Tavernarakis N, Kroemer G. Can autophagy promote longevity? Nat Cell Biol. 2010;12(9):842.
22. Warr MR, Binnewies M, Flach J, Reynaud D, Garg T, Malhotra R, Debnath J, Passegué E. FOXO3A directs a protective autophagy program in haematopoietic stem cells. Nature. 2013;494(7437):323–7.
23. Leveque-El Mouttie L, Vu T, Lineburg KE, Kuns RD, Bagger FO, Teal BE, Lor M, Boyle GM, Bruedigam C, Mintern JD, Hill GR. Autophagy is required for stem cell mobilization by G-CSF. Blood. 2015;125(19):2933–6.
24. Guan JL, Simon AK, Prescott M, Menendez JA, Liu F, Wang F, Wang C, Wolvetang E, Vazquez-Martin A, Zhang J. Autophagy in stem cells. Autophagy. 2013;9(6):830–49.
25. Phadwal K, Watson AS, Simon AK. Tightrope act: autophagy in stem cell renewal, differentiation, proliferation, and aging. Cell Mol Life Sci. 2013;70(1):89–103.
26. Vessoni AT, Filippi-Chiela EC, Menck CF, Lenz G. Autophagy and genomic integrity. Cell Death Differ. 2013;20(11):1444.
27. Pua HH, Dzhagalov I, Chuck M, Mizushima N, He YW. A critical role for the autophagy gene Atg5 in T cell survival and proliferation. J Exp Med. 2007;204(1):25–31.
28. Joshi A, Kundu M. Mitophagy in hematopoietic stem cells: the case for exploration. Autophagy. 2013;9(11):1737–49.
29. Hsu P, Qu CK. Metabolic plasticity and hematopoietic stem cell biology. Curr Opin Hematol. 2013;20(4):289.
30. Bowie MB, Kent DG, Dykstra B, McKnight KD, McCaffrey L, Hoodless PA, Eaves CJ. Identification of a new intrinsically timed developmental checkpoint that reprograms key hematopoietic stem cell properties. Proc Natl Acad Sci. 2007;104(14):5878–82.

31. Manesia JK, Xu Z, Broekaert D, Boon R, van Vliet A, Eelen G, Vanwelden T, Stegen S, Van Gastel N, Pascual-Montano A, Fendt SM. Highly proliferative primitive fetal liver hematopoietic stem cells are fueled by oxidative metabolic pathways. Stem Cell Res. 2015;15(3):715–21.
32. Vannini N, Girotra M, Naveiras O, Nikitin G, Campos V, Giger S, Roch A, Auwerx J, Lutolf MP. Specification of haematopoietic stem cell fate via modulation of mitochondrial activity. Nat Commun. 2016;7:13125.
33. Shaw AC, Joshi S, Greenwood H, Panda A, Lord JM. Aging of the innate immune system. Curr Opin Immunol. 2010;22(4):507–13.
34. Rossi DJ, Bryder D, Seita J, Nussenzweig A, Hoeijmakers J, Weissman IL. Deficiencies in DNA damage repair limit the function of haematopoietic stem cells with age. Nature. 2007;447(7145):725.
35. deHaan G, Van Zant G. Dynamic changes in mouse hematopoietic stem cell numbers during aging. Blood. 1999;93(10):3294–301.
36. Chambers SM, Shaw CA, Gatza C, Fisk CJ, Donehower LA, Goodell MA. Aging hematopoietic stem cells decline in function and exhibit epigenetic dysregulation. PLoS Biol. 2007;5(8):e201.
37. Wang J, Lu X, Sakk V, Klein CA, Rudolph KL. Senescence and apoptosis block hematopoietic activation of quiescent hematopoietic stem cells with short telomeres. Blood. 2014;124(22):3237–40.
38. Pearce DJ, Anjos-Afonso F, Ridler CM, Eddaoudi A, Bonnet D. Age-dependent increase in side population distribution within hematopoiesis: implications for our understanding of the mechanism of aging. Stem Cells. 2007;25(4):828–35.
39. Nguyen-McCarty M, Klein PS. Autophagy is a signature of a signaling network that maintains hematopoietic stem cells. PLoS One. 2017;12(5):e0177054.
40. deHaan G, Lazare S. Aging of hematopoietic stem cells. Blood. 2018;131:479.
41. Ito K, Hirao A, Arai F, Takubo K, Matsuoka S, Miyamoto K, Ohmura M, Naka K, Hosokawa K, Ikeda Y, Suda T. Reactive oxygen species act through p38 MAPK to limit the lifespan of hematopoietic stem cells. Nat Med. 2006;12(4):446.
42. Mohrin M, Shin J, Liu Y, Brown K, Luo H, Xi Y, Haynes CM, Chen D. A mitochondrial UPR-mediated metabolic checkpoint regulates hematopoietic stem cell aging. Science. 2015;347(6228):1374–7.
43. Norddahl GL, Pronk CJ, Wahlestedt M, Sten G, Nygren JM, Ugale A, Sigvardsson M, Bryder D. Accumulating mitochondrial DNA mutations drive premature hematopoietic aging phenotypes distinct from physiological stem cell aging. Cell Stem Cell. 2011;8(5):499–510.
44. Stolzing A, Jones E, McGonagle D, Scutt A. Age-related changes in human bone marrow-derived mesenchymal stem cells: consequences for cell therapies. Mech Ageing Dev. 2008;129(3):163–73.
45. Ishikawa ET, Gonzalez-Nieto D, Ghiaur G, Dunn SK, Ficker AM, Murali B, Madhu M, Gutstein DE, Fishman GI, Barrio LC, Cancelas JA. Connexin-43 prevents hematopoietic stem cell senescence through transfer of reactive oxygen species to bone marrow stromal cells. Proc Natl Acad Sci. 2012;109(23):9071–6.
46. Asumda FZ, Chase PB. Age-related changes in rat bone-marrow mesenchymal stem cell plasticity. BMC Cell Biol. 2011;12(1):44.
47. Singh S, Ghode S, Devi MR, Limaye L, Kale V. Phenotypic and functional characterization of a marrow-derived stromal cell line, M210B4 and its comparison with primary marrow stromal cells. Biomed Res J. 2015;2(1):120–33.
48. Tuljapurkar SR, McGuire TR, Brusnahan SK, Jackson JD, Garvin KL, Kessinger MA, Lane JT, O'Kane BJ, Sharp JG. Changes in human bone marrow fat content associated with changes in hematopoietic stem cell numbers and cytokine levels with aging. J Anat. 2011;219(5):574–81.
49. Naveiras O, Nardi V, Wenzel PL, Hauschka PV, Fahey F, Daley GQ. Bone-marrow adipocytes as negative regulators of the haematopoietic microenvironment. Nature. 2009;460(7252):259–63.
50. Kennedy DE, Knight KL. Inhibition of b lymphopoiesis by adipocytes and il-1–producing myeloid-derived suppressor cells. J Immunol. 2015;195(6):2666–74.

51. Bellantuono I, Aldahmash A, Kassem M. Aging of marrow stromal (skeletal) stem cells and their contribution to age-related bone loss. Biochim Biophys Acta. 2009;1792(4):364–70.
52. Freemont AJ, Hoyland JA. Morphology, mechanisms and pathology of musculoskeletal ageing. J Pathol. 2007;211(2):252–9.
53. Wagner W, Horn P, Bork S, Ho AD. Aging of hematopoietic stem cells is regulated by the stem cell niche. Exp Gerontol. 2008;43(11):974–80.
54. Pennock R, Bray E, Pryor P, James S, McKeegan P, Sturmey R, Genever P. Human cell dedifferentiation in mesenchymal condensates through controlled autophagy. Sci Rep. 2015;5:13113.
55. Kasper G, Mao L, Geissler S, Draycheva A, Trippens J, Kühnisch J, Tschirschmann M, Kaspar K, Perka C, Duda GN, Klose J. Insights into mesenchymal stem cell aging: involvement of antioxidant defense and actin cytoskeleton. Stem Cells. 2009;27(6):1288–97.
56. Oliver L, Hue E, Priault M, Vallette FM. Basal autophagy decreased during the differentiation of human adult mesenchymal stem cells. Stem Cells Dev. 2012;21(15):2779–88.
57. Li J, Zhou J, Zhang D, Song Y, She J, Bai C. Bone marrow-derived mesenchymal stem cells enhance autophagy via PI3K/AKT signalling to reduce the severity of ischaemia/reperfusion-induced lung injury. J Cell Mol Med. 2015;19(10):2341–51.
58. Zhang Q, Yang YJ, Wang H, Dong QT, Wang TJ, Qian HY, Xu H. Autophagy activation: a novel mechanism of atorvastatin to protect mesenchymal stem cells from hypoxia and serum deprivation via AMP-activated protein kinase/mammalian target of rapamycin pathway. Stem Cells Dev. 2012;21(8):1321–32.
59. Pantovic A, Krstic A, Janjetovic K, Kocic J, Harhaji-Trajkovic L, Bugarski D, Trajkovic V. Coordinated time-dependent modulation of AMPK/Akt/mTORsignalingand autophagy controls osteogenic differentiation of human mesenchymal stem cells. Bone. 2013;52(1):524–31.
60. Ma Y, Qi M, An Y, Zhang L, Yang R, Doro DH, Liu W, Jin Y. Autophagy controls mesenchymal stem cell properties and senescence during bone aging. Aging Cell. 2018;17(1):e12709.
61. Li L, Li L, Zhang Z, Jiang Z. Hypoxia promotes bone marrow-derived mesenchymal stem cell proliferation through apelin/APJ/autophagy pathway. Acta Biochim Biophys Sin. 2015;47(5):362–7.
62. Herberg S, Shi X, Johnson MH, Hamrick MW, Isales CM, Hill WD. Stromal cell-derived factor-1β mediates cell survival through enhancing autophagy in bone marrow-derived mesenchymal stem cells. PLoS One. 2013;8(3):e58207.
63. Hou J, Han ZP, Jing YY, Yang X, Zhang SS, Sun K, Hao C, Meng Y, Yu FH, Liu XQ, Shi YF. Autophagy prevents irradiation injury and maintains stemness through decreasing ROS generation in mesenchymal stem cells. Cell Death Dis. 2013;4(10):e844.
64. Jung J, Choi JH, Lee Y, Park JW, Oh IH, Hwang SG, Kim KS, Kim GJ. Human placenta-derived mesenchymal stem cells promote hepatic regeneration in CCl4-injured rat liver model via increased autophagic mechanism. Stem Cells. 2013;31(8):1584–96.
65. Gharibi B, Farzadi S, Ghuman M, Hughes FJ. Inhibition of Akt/mTOR attenuates age-related changes in mesenchymal stem cells. Stem Cells. 2014;32(8):2256–66.
66. Liu J, Hao H, Huang H, Tong C, Ti D, Dong L, Chen D, Zhao Y, Liu H, Han W, Fu X. Hypoxia regulates the therapeutic potential of mesenchymal stem cells through enhanced autophagy. Int J Low Extrem Wounds. 2015;14(1):63–72.
67. Zhao K, Hao H, Liu J, Tong C, Cheng Y, Xie Z, Zang L, Mu Y, Han W. Bone marrow-derived mesenchymal stem cells ameliorate chronic high glucose-induced β-cell injury through modulation of autophagy. Cell Death Dis. 2015;6(9):e1885.
68. Méndez-Ferrer S, Michurina TV, Ferraro F, Mazloom AR, MacArthur BD, Lira SA, Scadden DT, Ma'ayan A, Enikolopov GN, Frenette PS. Mesenchymal and haematopoietic stem cells form a unique bone marrow niche. Nature. 2010;466(7308):829.
69. He C, Klionsky DJ. Regulation mechanisms and signaling pathways of autophagy. Annu Rev Genet. 2009;43:67.
70. Takahashi K, Yamanaka S. Induction of pluripotent stem cells from mouse embryonic and adult fibroblast cultures by defined factors. Cell. 2006;126(4):663–76.

71. Wahlestedt M, Norddahl GL, Sten G, Ugale A, Frisk MA, Mattsson R, Deierborg T, Sigvardsson M, Bryder D. An epigenetic component of hematopoietic stem cell aging amenable to reprogramming into a young state. Blood. 2013;121(21):4257–64.
72. Satoh Y, Yokota T, Sudo T, Kondo M, Lai A, Kincade PW, Kouro T, Iida R, Kokame K, Miyata T, Habuchi Y. The Satb1 protein directs hematopoietic stem cell differentiation toward lymphoid lineages. Immunity. 2013;38(6):1105–15.
73. Wang J, Morita Y, Han B, Niemann S, Löffler B, Rudolph KL. Per2 induction limits lymphoid-biased haematopoietic stem cells and lymphopoiesis in the context of DNA damage and ageing. Nat Cell Biol. 2016;18(5):480–90.
74. Brown K, Xie S, Qiu X, Mohrin M, Shin J, Liu Y, Zhang D, Scadden DT, Chen D. SIRT3 reverses aging-associated degeneration. Cell Rep. 2013;3(2):319–27.
75. Lamming DW, Ye L, Sabatini DM, Baur JA. Rapalogs and mTOR inhibitors as anti-aging therapeutics. J Clin Investig. 2013;123(3):980–9.
76. Chen C, Liu Y, Liu Y, Zheng P. mTOR regulation and therapeutic rejuvenation of aging hematopoietic stem cells. Sci Signal. 2009;2(98):ra75.
77. Cheng CW, Adams GB, Perin L, Wei M, Zhou X, Lam BS, Da Sacco S, Mirisola M, Quinn DI, Dorff TB, Kopchick JJ. Prolonged fasting reduces IGF-1/PKA to promote hematopoietic-stem-cell-based regeneration and reverse immunosuppression. Cell Stem Cell. 2014;14(6):810–23.
78. Chen J, Astle CM, Harrison DE. Hematopoietic senescence is postponed and hematopoietic stem cell function is enhanced by dietary restriction. Exp Hematol. 2003;31(11):1097–103.
79. Florian MC, Nattamai KJ, Dörr K, Marka G, Überle B, Vas V, Eckl C, Andrä I, Schiemann M, Oostendorp RA, Scharffetter-Kochanek K. A canonical to non-canonical Wnt signalling switch in haematopoietic stem-cell ageing. Nature. 2013;503(7476):392–6.
80. Florian MC, Dörr K, Niebel A, Daria D, Schrezenmeier H, Rojewski M, Filippi MD, Hasenberg A, Gunzer M, Scharffetter-Kochanek K, Zheng Y. Cdc42 activity regulates hematopoietic stem cell aging and rejuvenation. Cell Stem Cell. 2012;10(5):520–30.
81. Chang J, Wang Y, Shao L, Laberge RM, Demaria M, Campisi J, Janakiraman K, Sharpless NE, Ding S, Feng W, Luo Y. Clearance of senescent cells by ABT263 rejuvenates aged hematopoietic stem cells in mice. Nat Med. 2016;22(1):78–83.
82. Galipeau J, Sensébé L. Mesenchymal stromal cells: clinical challenges and therapeutic opportunities. Cell Stem Cell. 2018;22(6):824–33.
83. Singh S, Moirangthem RD, Vaidya A, Jalnapurkar S, Limaye L, Kale V. AKT signaling prevailing in mesenchymal stromal cells modulates the functionality of hematopoietic stem cells via intercellular communication. Stem Cells. 2016;34(9):2354–67.

# Chapter 4
# Forkhead Box O (FoxO) Transcription Factors in Autophagy, Metabolic Health, and Tissue Homeostasis

**Longhua Liu and Zhiyong Cheng**

**Abstract** As transcription factors, the forkhead box O family proteins control the expression of genes that are involved in the regulation of autophagy and metabolism. The FoxO–autophagy axis has been shown to mediate cell differentiation and tissue development. Dysregulated FoxO activity may compromise tissue development and homeostasis, concomitant with metabolic abnormalities across tissues such as liver, adipose tissue, skeletal muscle, and heart. In this chapter, we discuss the mechanism or pathways of FoxO transcription factors regulating autophagy and tissue integrity, and the FoxO–autophagy axis in cellular metabolism and fate determination. The evidence summarized here suggests that targeting the FoxO–autophagy axis may lead to therapeutic options for metabolic derangements and cell or tissue dysfunction.

## Abbreviations

| | |
|---|---|
| ACO | Acyl-CoA oxidase |
| Akt (or PKB) | Protein kinase B |
| AMPK | AMP-activated protein kinase |
| Atg | Autophagy related protein |
| C/EBP | CCAAT/enhancer-binding protein |
| CD36 | Fatty acid translocase FAT/cluster of differentiation 36 |
| CMV | Controlled mechanical ventilation |

L. Liu
Department of Human Nutrition, Foods, and Exercise, Fralin Life Science Institute, College of Agriculture and Life Science, Virginia Tech, Blacksburg, VA, USA

Department of Pathology & Cell Biology, Columbia Diabetes Research Center, New York Obesity Research Center, Columbia University, New York, NY, USA

Z. Cheng (✉)
Department of Human Nutrition, Foods, and Exercise, Fralin Life Science Institute, College of Agriculture and Life Science, Virginia Tech, Blacksburg, VA, USA
e-mail: zcheng@vt.edu



K. Turksen (ed.), *Autophagy in Health and Disease*, Stem Cell Biology and Regenerative Medicine, https://doi.org/10.1007/978-3-319-98146-8_4

| | |
|---|---|
| CREB | cAMP response element binding protein |
| DBD | DNA binding domain |
| ER | Endoplasmic reticulum |
| FoxO | Forkhead box O |
| FSP27 | Fat-specific protein 27 |
| G6Pase | Glucose-6-phosphatase |
| HDAC | Histone deacetylase |
| HK | Hexokinase |
| JNK | c-Jun N-terminal kinase |
| KAA | Ketogenic amino acid |
| LC3 | Microtubule-associated protein 1A/1B-light chain 3-phosphatidylethanolamine conjugate |
| LDHA | Lactate dehydrogenase A |
| LXR | Liver X receptor |
| MI | Myocardial infarction |
| MST1 | Mammalian sterile 20-like kinase 1 |
| mTORC1 | Mammalian target of rapamycin complex 1 |
| MTP | Microsomal tryglyceride transfer protein |
| MuRF1 | Muscle RING-finger protein-1 |
| NES | Nuclear export sequence |
| NLS | Nuclear localization signal |
| Pdx1 | Pancreas/duodenum homeobox gene-1 |
| PEPCK | Phosphoenolpyruvate carboxykinase |
| PGC1 | Peroxisome proliferator-activated receptor gamma coactivator 1 |
| PI3K | Phosphatidylinositol 3 kinase |
| PI3P | Phoshpatidylinositol 3-phosphate |
| PKA | Protein kinase A |
| PKM2 | Pyruvate kinase isozymes M2 |
| PPARγ | Peroxisome proliferator-activated receptor γ |
| RXR | Retinoid X receptor |
| Sirt1,2 | sirtuin 1, 2 |
| SKP2 | S-phase kinase-associated protein 2 |
| SQSTM1 (or p62) | Sequestosome 1 |
| SREBP | Sterol response element-binding protein |
| STZ | Streptozotocin |
| Tfeb | Transcription factor EB |
| ULK | Unc-51-like kinase |
| VLDL | Very-low-density lipoprotein |
| Vps34 | Vacuolar proteins 34 |

## 4.1 The FoxO Family

The forkhead box O (FoxO) family is a subclass of winged helix/forkhead transcription factors. The mammalian FoxO family are homologues of the *Caenorhabditis elegans* transcription factor Dauer Formation-16 (DAF-16), including four members (FoxO1, FoxO3, FoxO4, and FoxO6) [1–5]. FoxO proteins consist of four domains, including a highly conserved DNA binding domain (DBD), a nuclear localization signal (NLS) domain, a nuclear export sequence (NES) domain, and a C-terminal transactivation domain [6, 7]. The activities of FoxO transcription factors can be regulated by posttranslational modifications, such as phosphorylation, acetylation, and ubiquitination [7, 8]. For instance, insulin activates phosphatidylinositol 3 kinase (PI3K)-Akt/protein kinase B (PKB) pathway, where Akt/PKB (a serine–threonine kinase) can phosphorylate FoxOs and exclude them from nucleus, thereby inhibiting the transcriptional activity [7–9]. Akt-mediated phosphorylation of FoxO1 promotes cytoplasmic retention of FoxO1 and ubiquitination, eventually leading to proteasomal degradation [10]. FoxO1 ubiquitination and degradation can also be stimulated by S-phase kinase-associated protein 2 (SKP2), a subunit of the Skp1–Cul1–F-box protein ubiquitin complex [11]. Deacetylation of FoxO3 by Sirt1 seems to differentially regulate FoxO3 activities, promoting FoxO3-meidated cell cycle arrest and resistance to cell stress but dampening FoxO3-mediated cell death [12]. Accumulating evidence suggests that FoxO proteins play an important role in glucose and lipid metabolism in liver, muscle and adipose tissue [13–18]. In addition, metabolic changes in obesity, diabetes, and cardiometabolic diseases have been linked to aberrant autophagy, where FoxO transcription factors regulate an array of autophagy genes [19–26]. These studies shed new light on the mechanisms or pathways by which FoxO proteins regulate metabolism, and underscore FoxO family as potential targets to treat metabolic diseases. In this chapter, we review the evidence of FoxO transcription factors in the regulation of autophagy, metabolic health, and tissue homeostasis.

## 4.2 FoxOs in Metabolic Health

### *4.2.1 Adipose Tissue*

Adipose tissue plays an important role in energy metabolism by storing extra energy as fat and by secreting various adipokines (e.g., adiponectin and leptin) to regulate systemic metabolism [27, 28]. Studies of FoxOs, particularly the most abundant isoform FoxO1 in adipose tissue and adipogenesis have led to two mechanistic models (Fig. 4.1). Earlier studies suggested that FoxO1 was a suppressor of adipogenesis (Fig. 4.1a). Overexpression of a constitutively active form of FoxO1 was reported to inhibit differentiation of preadipocytes, while dominant-negative FoxO1 rescues the differentiation [14]. FoxO1 haploinsufficiency can protect mice from

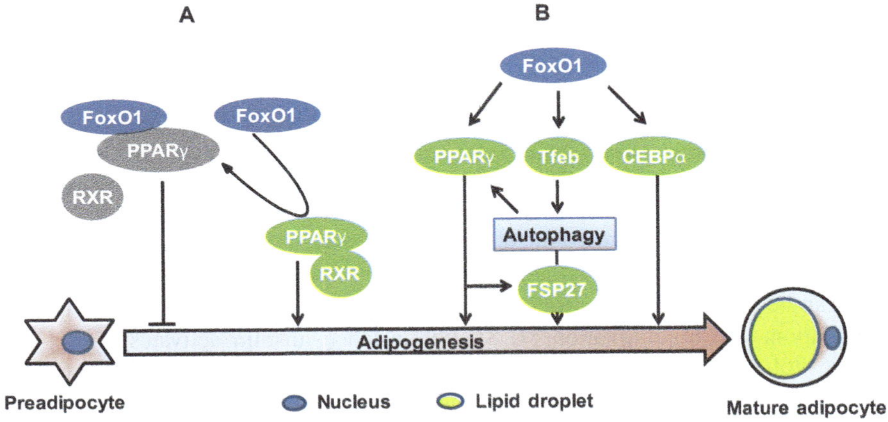


**Fig. 4.1** The role of FoxO1 in adipogenesis or adipocyte differentiation. Two models have been proposed: (**a**) PPARγ is a major positive regulator of adipogenesis, while FoxO1 transrepresses PPARγ and forms complex with PPARγ to inhibit adipogenesis. (**b**) FoxO1 promotes adipogenesis via the induction of PPARγ and CEBPα, or Tfeb that induces autophagy to stabilize PPARγ and FSP27, thereby enhancing lipid accumulation and adipocyte differentiation. See the text in Sects. 4.2.1 and 4.3.2 for detailed discussion. *FoxO1* factor forkhead box O1, *C/EBP* CCAAT/enhancer-binding protein, *PPARγ* peroxisome proliferator-activated receptor gamma, *RXR* retinoid X receptor, *Tfeb* transcription factor EB

diet-induced diabetes [14]. Further study showed that complex of PPARγ and RXRα was required to promote adiogenesis, while FoxO1 tended to disrupt the complex of PPARγ and RXRα and inhibit adipocyte differentiation [29]. FoxO1 may also transrepress PPARγ by direct protein–protein interaction [30], or inhibit the expression of PPARγ1 and PPARγ2 in rat primary adipocytes and increase the expression of Glut 4, thereby improving insulin sensitivity [31]. Overexpression of a mutant FoxO1 in adipose tissue-specific FoxO1 transgenic mice can increase Glut 4 and improve glucose tolerance and insulin sensitivity [32]. Posttranslational modification, e.g., acetylation of Foxo1, also affects the FoxO1 role in adipocyte differentiation. Overexpression of Sirt2, which is a cytoplasmic predominant sirtuin in adipocytes, decreases the acetylation of FoxO1 and reduces the expression of PPARγ, CEBPα and other genes involved in terminal adipocyte differentiation, resulting in the inhibition of differentiation; by contrast, knockdown of Sirt2 promotes adipogenesis [33]. Resveratrol increases Sirt1, but decreases FoxO1 and PPARγ2 and inhibits differentiation. However, nicotinamide can decrease Sirt1 mRNA but increase FoxO1 and PPARγ2 to stimulate the differentiation in pig preadipocytes [34]. Recent studies, nevertheless, suggest that FoxO1 is an enhancer of adipocyte differentiation [25, 26, 35–38] (Fig. 4.1b). Knockdown of FoxO1 using adenovirus FoxO1-siRNA inhibits the differentiation of 3T3-L1 preadipocyte and is accompanied by decreased PPARγ and CEBPα, especially when exposing cells to FoxO1-siRNA before the differentiation induction [35, 37]. Epigallocatechin gallate (EGCG) treatment can inhibit the adipocyte differentiation via inactivation of

FoxO1 [36]. Mechanistic studies underscore autophagy as an important regulator of adipocyte differentiation, presumably through stabilization of PPARγ and induction of FSP27 [25, 26]. The role of FoxO1 in adipocyte autophagy and biology will be discussed in more detail in Sect. 4.3.

### *4.2.2 Liver*

The liver is a major organ for glucose and lipid metabolism (Fig. 4.2). Plasma glucose levels are maintained well through glucose production and uptake by peripheral tissues such as the liver, adipose tissue, and skeletal muscle in normal conditions [39]. During a fasting period, the liver will break down glycogen and increase gluconeogenesis to increase the glucose levels; while under postprandial condition, the liver can store glucose through glycogen synthesis and reduce glucose level in the bloodstream [8]. Glucose-6-phosphatase (G6Pase), fructose-1, 6-bisphosphatase, and phosphoenolpyruvate carboxykinase (PEPCK) are three major gluconeogenic enzymes in liver. Overexpression of a constitutively active form of FoxO1 in the

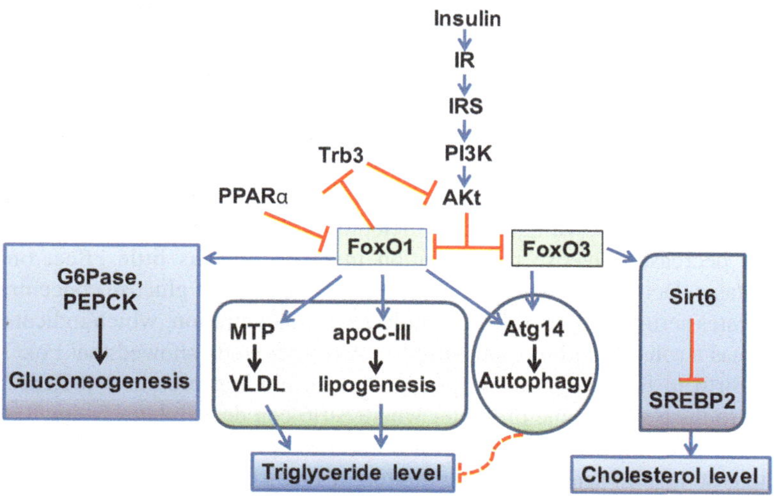


**Fig. 4.2** FoxOs regulate autophagy and metabolic homeostasis in the liver. Under postprandial conditions, insulin signaling pathway is activated to inhibit FoxO1 and FoxO3. In fasting state, FoxO1 is activated and can increase gluconeogenesis through G6Pase and PEPCK. FoxO1 can activate MTP/VLDL pathway and elevate lipogenesis through apoC-III to increase triglyceride level in serum. Interestingly, FoxOs were shown to regulate autophagy via Atg14 to decrease triglyceride level in serum. FoxO3 can decrease cholesterol level through Sirt6 and further inhibition of SREBP2. Moreover, FoxO1 may enhance the insulin activity by inhibiting Trb3 which prevents the phosphorylation of Akt. See the text in Sects. 4.2.2 and 4.3.3 for detailed discussion. *Atg14* autophagy related 14, *G6Pase* Glucose-6-phosphatase, *MTP* microsomal triglyceride transfer protein, *PEPCK* phosphoenolpyruvate carboxykinase, *Sirt6* Sirtuin 6, *SREBP2* sterol response element-binding protein 2, *Trb3* tribbles homolog 3, *VLDL* very-low-density lipoprotein

liver of transgenic mice increases fasting glucose levels and glucose intolerance. The genes involved in gluconeogenesis, such as G6Pase and PEPCK, are also increased; while de novo lipogenesis is decreased after refeeding [15]. Inhibition of FoxO1 expression using antisense oligonucleotides (ASOs) in mouse hepatocytes can decrease mRNA levels of G6Pase and PEPCK and lipolysis, while increasing insulin sensitivity and glucose tolerance in mice with diet-induced obesity [40]. Recently, similar results have shown that liver-specific ablation of insulin receptor (LIRKO) increases glucose intolerance and insulin resistance, while liver-specific double knockout insulin receptor and FoxO1 (LIRFKO) rescue glucose tolerance, reduce hepatic glucose production, and decrease gluconeogenic gene expression, such as G6Pase and PEPCK [41]. The abovementioned data show that FoxO1 can increase hepatic glucose production through upregulating G6Pase and PEPCK.

FoxOs also play roles in lipid metabolism. Adenoviral delivery of FoxO1 to mouse liver can induce steatosis through increasing triglyceride accumulation but decreasing fatty acid oxidation [42]. Another study showed that in HepG2 cells, FoxO1 can induce the expression of microsomal tryglyceride transfer protein (MTP), by directly binding to the MTP promoter [43]. MTP is a rate-limiting protein involved in very-low-density lipoprotein (VLDL) production. Overexpression of a constitutively active FoxO1 in mice can incease the expression of MTP, VLDL production, and plasma triglyceride levels [43]. Adenovirus-mediated FoxO1 delivered to hepatocytes stimulates the expression of apoC-III through directly binding to the promoter of apoC-III, which plays an important role in triglyceride metabolism, and results in increased triglyceride levels in the plasma and fat intolerance [44]. A mutation or deletion of the FoxO1 binding site on the apoC-III promoter can abolish the insulin response and FoxO1-mediated stimulation. A further study showed that PPARα interacts with FoxO1 to regulate the apoC-III expression [45]. Intrestingly, it was shown that FoxO1 deletion in the liver using Cre/LoxP genetic approch decreases glucose concentration in blood but has little effect on lipid homeostasis. Deletion of both FoxO1 and FoxO3 decreases glucose concentrations and elevates serum tryglyceride and cholesterol concentration, which indicates that FoxO3 has a role in lipid metabolism [46]. A further study showed that FoxO3 can recruit Sirt6 to regulate sterol response element-binding protein 2 (SREBP2) by binding to the Srebp2 gene promoter where Sirt6 can deacetylates lysine 9 and 56 on histone H3, and further cholesterol homestasis [47]. It was shown that FoxO1 can increase insulin sensitivity by inhibiting tribbles homolog 3(Trb3), which prevents the phosphorylation of Akt by binding to it, to increase the phosphorylation of Akt [42]. Above data show the important roles of FoxOs in the regulation of glucose and lipid metabolism.

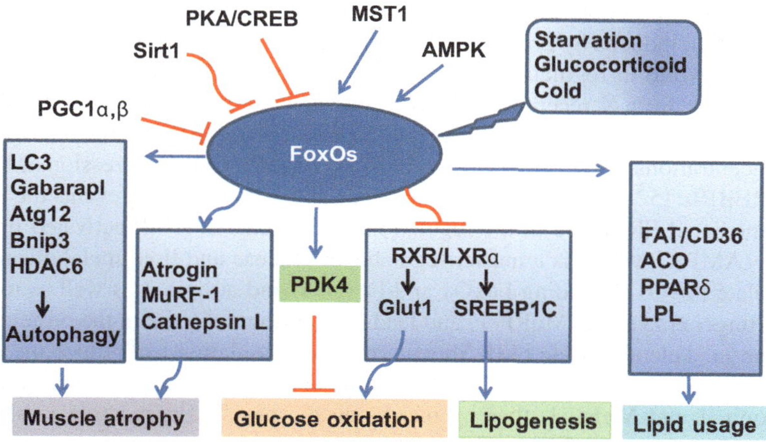


**Fig. 4.3** FoxOs regulate autophagy, metabolism, and tissue homeostasis in skeletal muscle. Under starvation, denervation, or cold conditions, FoxOs are activated and switch substrate metabolism from glucose oxidation to fatty acid utilization. When FoxOs are activated, they inhibit glucose oxidation by promoting PDK4 and disrupting RXR–LXRα complex, which also leads to decreased lipogenesis through SREBP1C. FoxOs can increase lipid usage through activating FAT, CD36, ACO, PPARδ, and LPL. Not only can FoxOs activate autophagy via LC3, Gabarapl, Atg12, Bnip3, and HDAC6, but also they can promote ubiquitin–proteasome pathway through atrogin, MuRF-1, and cathepsin L. Activation of autophagy and ubiquitin–proteasome pathways by FoxOs causes muscle atrophy. Other factors such as AMPK and MST1 can also activate FoxOs, while PGC1α, PGC1β, Sirt1, PKA, and CREB inhibit the activity of FoxOs. See the text in Sects. 4.2.3 and 4.3.4 for detailed discussion. *ACO* Acyl-CoA oxidase, *AMPK* AMP-activated protein kinase, *CD36* cluster of differentiation 36, *FAT* fatty acid translocase, *HDAC6* histone deacetylase 6, *LPL* lipoprotein lipase, *LXRα* liver X receptor α, *MST1* Mammalian Sterile 20-like kinase 1, *MuRF-1* muscle RING finger enzyme-1, *PDK4* pyruvate dehydrogenase kinase 4, *PKA* protein kinase A

## 4.2.3 Muscle

Skeletal muscle is a major peripheral tissue that is responsible for insulin-mediated energy homeostasis, which contributes more than 30% of resting metabolic rate and 80% whole body glucose uptake [48]. FoxO1 plays important roles in regulating glucose and lipid metabolism in skeletal muscle (Fig. 4.3). Under starvation or glucocorticoid treatment, FoxO1 is induced and pyruvate dehydrogenase kinase 4 (PDK4) is upregulated by FoxO1 which can directly bind to the promoter of PDK4 and further inhibit glucose oxidation [13]. A further study shows that the effect of RXRγ on enhanced glucose tolerance may be at least in part due to upregulated Glut1 in skeletal muscle [49]. Another study showed that ectopic expression of FoxO1 increased the gene expression of lipoprotein lipase (LPL), which plays an important role in lipid usage in skeletal muscle [50]. Overexpression of FoxO1 using inducible constructs in C2C12 cells increases the fatty acid translocase FAT/cluster of differentiation 36 (CD36), Acyl-CoA oxidase (ACO) and PPARδ, and enhances the uptake of oleate and oleate oxidation [51]. These effects of enhanced FA utilization induced by FoxO1 can be abolished by the CD36 inhibitor. SREBP1c also plays a role in regulation of lipid metabolism via FoxO1. RXRα or RXRγ, together with liver X receptor α (LXRα), can activate the promoter of SREBP1c [52]. Overexpression of RXRγ in skeletal muscle increases SREBP1c and triglyceride concentrations; and overexpression of FoxO1 decreases the expression of RXRγ and SREBP1c [52]. The above-mentioned data show that FoxO1 prevents the RXR/LXR-mediated SREBP1c in the regulation of lipogenesis. AMP-activated protein kinase (AMPK), which is a master regulator of glucose and lipid metabolism, can upregulate FoxOs (including FoxO1 and FoxO3a) and atrogin-1 as well as muscle RING finger enzyme-1 (MuRF-1) and leads to a further increase in the protein degradation in skeletal muscle [53]. Short-term cold stimulation can reduce the phosphorylation of Akt and increase the activity of FoxO1, which increases the expression of atrogin-1 and MuRF-1, thereby increasing the protein degradation in skeletal muscle [54]. Overexpression of FoxO1 in skeletal muscle can reduce the body weight and skeletal muscle mass, decrease type I muscles. Overexpression of FoxO1 can also increase the expression of cathepsin L, which is a lysosomal proteinase, leading to increased protein degradation in skeletal muscle [55]. Hence, not only do FoxOs play important roles in switching from glucose oxidation to fatty acid utilization, but they also increase muscle atrophy in skeletal muscle.

### 4.2.4 *Pancreatic β-Cells*

The pancreas plays a critical role in glucose homeostasis and contains at least five different types of cells: α-cells, β-cells, δ-cells, ε-cells, and PP-cells [56]. β-cells can sense the blood glucose concentration and secrete insulin to regulate glucose homeostasis. Generally, β-cells can uptake and metabolize glucose, thereby increasing the cellular ratio of ATP/ADP to close the $K^+$-ATP channel and depolarize the cell. Depolarization of the cell leads to the opening of the voltage-dependent $Ca^{2+}$ channel and a subsequent increase the cytosolic $Ca^{2+}$ concentration, leading to the release of insulin [57]. Previous studies show that insulin signaling is required to maintain β-cell mass. Mice with insulin receptor substrate-2 knockout (IRS2KO) develop β-cell failure. While haploinsufficiency of FoxO1 can reverse β-cell failure in IRS2KO mice through the increase in β-cell proliferation as well as increase the expression of pancreas/duodenum homeobox gene-1 (Pdx1), which is a pancreatic transcription factor [58]. Overexpressing a constitutively active form of FoxO1 decreases the Pdx1 expression by acting as a repressor of Foxa2-dependent Pdx1 transcription. These data show that insulin signaling can increase the expression of Pdx1 to maintain β-cell mass through inhibition of FoxO1. A further study showed that the deletion of FoxO1 in the domain of the Pdx1 promoter (P-FoxO1-KO) improved glucose tolerance under high-fat high-sucrose diet (HFHSD). Additionally, this led to an increased β-cell mass. Mice with P-FoxO1-KO crossed with db/db showed more severe glucose intolerance than in the control mice, indicating that

FoxO1 functions as a double-edged sword in the pancreas [59]. FoxO1 can also regulate β-cell function, survival, and compensation through the inhibition of PPARγ and its target genes, such as Pdx1 and pyruvate carboxylase [60]. Impaired FoxO1 may cause or exacerbate diabetes [60]. Controversially, upregulated FoxO1 in β-cell-specific FoxO1-transgenic mice can increase β-cell mass, improve glucose tolerance, and protect the mice from HFD-induced glucose disorder [61]. Furthermore, FoxO1 plays other roles in β-cell function. It was shown that FoxO1 could protect β-cell from oxidative stress by increasing two insulin2 gene transcription factors, NeuroD and MafA, through forming a complex with promyelocytic leukemia protein (Pml) and Sirt1. Acetylated FoxO1 can bind to Pml and prevent FoxO1 from degradation. While Sirt1 can deacetylate FoxO1 and accelerate its degradation, thereby prevent unchecked FoxO1 transcription [62]. Another study shows that the c-Jun N-terminal kinase (JNK) pathway can regulate FoxO1 translocation from the cytosol to the nucleus. Overexpression of JNK can induce the nuclear localization of FoxO1, while inhibiting JNK can decrease the oxidative stress-induced FoxO1 nuclear localization [63]. Further studies are warrented to elucidate the complicated roles of FoxO1 in β-cell function, proliferation, and compensation, especially under different physiological conditions, such as diabetes.

## 4.3 FoxOs in Autophagy and Tissue Homeostasis

### *4.3.1 Autophagy and FoxO Network*

Autophagy (from the Latin words "auto," means oneself and "phagy," meaning to eat) refers to physiological degradative processes during which cytosolic components are degraded in bulk. This includes the degradation of proteins, lipids, sugars, and some organelles (mitochondria, peroxisomes, ribosomes) [64]. There are three types of autophagy, including microautophagy, macroautophagy, and chaperone-mediated autophagy. Microautophagy is the process where the lysosome membrane engulfs part of cytoplasm, and macroautophagy is the degradative process that involves a formation of autophagosome, while chaperone-mediated autophagy is a selective degradative process that can specifically degrade proteins recognized by the chaperone protein Hsc70. Increasing evidence has demonstrated that FoxO transcription factors regulate macroautophagy (referred to as autophagy hereafter) in different stages of this process (Fig. 4.4).

Autophagy is a well-regulated degradative system and many factors are involved in this process, which can be divided into several steps: initiation, elongation, maturation, autophagosome–lysosome fusion, and degradation (Fig. 4.4) [64]. Although the sources of isolation membranes are still not fully understood, endoplasmic reticulum (ER) is the most important isolation membrane contributor to initiate the autophagy under starvation. The Golgi complex, endosome, mitochondria, plasma membrane, and nuclear membrane are also possible membrane sources [65]. Under

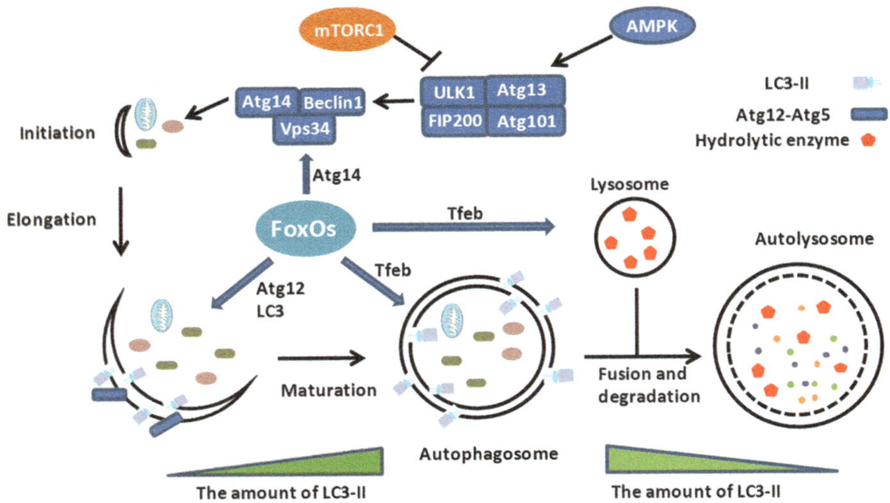


**Fig. 4.4** The general autophagy pathway and FoxO network. Autophagy (macroautophagy) is a physiological degradative process, which can degrade cytosolic components (such as dysfunctional proteins and organelles). Autophagy can be generally divided into several steps: initiation, elongation, maturation, fusion, and degradation. During nutrient sufficiency, mTORC1 can inhibit the activity of ULK1 complex to prevent the induction of autophagy. While under starvation conditions, AMPK can activate ULK1 complex to initiate the autophagy through recruiting Atg14–beclin1–Vps34 complex. The Atg12–Atg5 complex and LC3-II are important for the elongation and maturation of autophagy, during which the amount of LC3-II is gradually increased. At the end of maturation, autophagosome is formed and then fuse with lysosome to form autolysosome. In the autolysosome, the engulfed cytosolic components including part of intra-autophagosome located LC3-II can be degraded. So, during the degradation, the amount of LC3-II is decreased. FoxO transcription factors have been shown to regulate multiple stages of autophagy by targeting Atg14, Atg12, LC3, and Tfeb. See the text in Sect. 4.3.1 for detailed discussion. *mTORC1* mammalian target of rapamycin complex 1, *Tfeb* transcriptionfactor EB, *ULK1* Unc-51-like kinase 1

starvation, following the induction of autophagy, a Ω-like shape domain on ER is formed which termed as omegasome. Phoshpatidylinositol 3-phosphate (PI3P), autophagy related protein 14 L (Atg14L), beclin1, vacuolar proteins 34(Vps34) and Unc-51-like kinase 1(ULK1) are required for the formation of omegasome [66–71]. Among many factors involved in the initiation of autophagy, Atg1–Unc-51-like kinase (ULK) complex plays a central role. In mammals, ULK1–atg13–the focal adhesion kinase family interacting protein of 200 kDa (FIP200)–atg101 kinase complex is negatively regulated by the mammalian target of rapamycin complex 1 (mTORC1) depending on nutrient status. For instance, under nutrient sufficiency, high mTOR activity can phosphorylate ULK1 at Ser757 to prevent ULK1 activity and disrupt the ULK1–AMPK interaction, resulting in inhibition of autophagy [72]. Under starvation, AMPK can phosphorylate ULK1 at Ser317 and Ser777 directly and activate ULK1 to induce autophagy. Ser467 and Ser 555 on ULK1 can be also phosphorylated by AMPK to promote autophagy [73].

Following the formation of the isolation membrane, it will elongate to engulf cytosol components (Fig. 4.4). During the elongation of autophagy, Atg12-Atg5 translocates to the outside of the isolation membrane and detaches from the membrane before or after the autophagosome is formed completely [74]. The Atg12–Atg5 complex is required for targeting microtubule-associated protein 1A/1B-light chain 3-phosphatidylethanolamine conjugate (LC3) (a mammalian homolog of Atg8) onto the isolation membrane. LC3 is synthesized as a preform of ProLC3, which can be cleaved by Atg4 to expose the C-terminal Gly of LC3 (LC3-I). LC3I can conjugate with PE to form LC3-II (LC3-PE) activated by Atg7 (E1-like enzyme) and Atg3 (E2-like enzyme) [75]. During the elongation to maturation of autophagosome, the amount of LC3-II is increased. Atg4 can then delipidate LC3-II to LC3-I on the surface of autohpagosome to recycle LC3-I. After the maturation of autophagosome, the outer membrane can fuse with the lysosome to form an autolysosome (Fig. 4.4). The process of fusion can be positively regulated by UVRAG–Vps34–beclin1 PI3K complex, but can be also negatively regulated by Rubicon–UVRAG–Vps34–beclin1 PI3K complex [66, 67]. The engulfed cytosolic components, including part of LC3-II which locates on the intra-autophagosome surface, can be degraded by hydrolases in the autolysosome. During the degradation process, the amount of LC3-II will be decreased. In addition to nonselective autophagy, selective autophagy is also physiologically important. P62/SQSTM1, an important ubiquitin and LC3-binding protein as an autophagy adaptor, can be selectively degraded by the autolysosome [76, 77]. When autophagy is impaired, P62 accumulates in the cell and serves as an important biomarker for autophagy. Based on the different organelles or cytosol components involved in selective degradation, the selective autophagy family includes mitophagy (specifically for mitochondria), pexophagy (for peroxisomes), and ribophagy (for ribosomes), during which P62 serves as a critical autophagy adaptor [78].

Emerging evidence suggests that FoxO proteins may interact with autophagy in different stages (Fig. 4.4). For instance, FoxOs can induce hepatic Atg14, which is critical for autophagy initiation [79]. As the mediators of elongation and maturation steps, LC3 and Atg12 can also be controlled by FoxOs directly in skeletal muscle and heart [21, 23, 80]. In adipocytes, FoxO transcription factor was shown to bind the promoter of Tfeb (transcription factor EB), the key regulator of autophagosome and lysosome [81, 82], and upregulate Tfeb expression [25]. Inhibition of FoxO transcription factor dampens Tfeb and autophagy in adipocytes [25]. The functional perspectives of FoxO–autophagy axis will be discussed in Sects. 4.3.2–4.3.6 below.

### 4.3.2 FoxO–Autophagy Axis in Adipocytes and Adipose Tissue

Adipocyte differentiation underpins adipose tissue development, in which autophagy plays a central role [83–85]. Deletion of autophagy gene (e.g., Atg5 or Atg7) suppresses adipocyte differentiation, compromising adipose tissue development and causing sudden death of mice at young ages [83–85]. FoxO1 may induce adipocyte autophagy through Tfeb (Fig. 4.1) [25, 26]. FoxO1 expression and activity is elevated during adipogenesis, which is paralleled with upregulation of Tfeb, the key regulator of autophagosome and lysosome. Indeed, autophagy activity is enhanced during adipocyte differentiation. ChIP assays confirm that FoxO1 can directly bind to the promoter of Tfeb [25]. Consistently, inhibitor of FoxO1 blocks the interaction between FoxO1 and Tfeb promoter, resulting in reduced Tfeb transcript and protein expression.

The FoxO1–autophagy axis plays a critical role in adipocyte biology [25, 26]. First, it underpins adipocyte differentiation and lipid droplet growth via FSP27. Pharmacological inhibition of FoxO1 (using FoxO1 specific inhibitor AS1842856) or autophagy (using autophagy inhibitor bafilomycin A1 plus leupeptin) similarly downregulates FSP27, which prevents lipid accumulation and adipocyte differentiation [25, 26, 38]. In mature (or terminally differentiated) adipocytes, blockage of the FoxO1–autophagy axis leads to smaller but more numerous lipid droplets, a phenotype frequently observed in browning of adipose tissue. Upregulation of FSP27 by the FoxO1–autophagy axis is likely to arise from the stabilizing effect on PPARγ, the key regulator of FSP27 expression and adipogenesis (Fig. 4.1) [85, 86]. Secondly, inhibition of the FoxO1–autophagy axis differentially regulates UCP1, UCP2, and UCP3 in adipocytes [25]. In particular, UCP1 is induced by the inhibition of FoxO1–autophagy axis, serving as a second line of evidence of browning of white adipocyte [87–89]. Given that positive energy balance contributes to obesity, targeting the FoxO1–autophagy axis may have the potential to treat or prevent obesity.

### 4.3.3 *FoxO–Autophagy Axis in the Liver*

FoxO1 plays an important role in hepatic lipid metabolism partly through the autophagy pathway (Fig. 4.2). FoxOs can regulate a key autophagy-related regulator, Atg14, through binding at its promoter, which has been revealed through luciferase reporter analysis and chromatin immunoprecipitation (ChIP) [79]. Either knockdown Atg14 or liver-specific FoxO1/3/4 triple knockout elevated triglycerides in the liver and serum [79]. Overexpression of Atg14 improved hepertriglyceridemia in liver-specific FoxO1/3/4 triple knockout mice [79]. A ketogenic amino acid (KAA) replacement diet ameliorated autophagy deficiency in HFD-fed mice accompanied by decreased FoxO3, increased Sirtuin 1(Sirt1), and inhibition of the phosphorylation of the mammalian target of rapamacin (mTOR) [90]. Metformin alleviated hepatosteatosis through Sirt1-mediated, AMP-activated kinase (AMPK)-independent autophagy machinery [91].

### 4.3.4 FoxO–Autophagy Axis in the Muscle

During fasting or denervation, FoxO3 can be activated to induce autophagy in skeletal muscle and cause muscle atrophy both in vitro and in vivo [21, 23] (Fig. 4.3). Phosphorylation of Akt can block FoxO3 to induce autophagy. However, rapamycin, which is a specific mTOR inhibitor, cannot rescue the effect of Akt on FoxO3 and autophagy, which indicates that FoxO3 can induce autophagy in an mTOR-independent way. FoxO3 regulates the expression of many autophagy-related genes, including LC3 and Bnip3 [21, 23]. Recently, triple FoxO1, 3, 4 muscle-specific knockout mice were generated to study the role of FoxOs in autophagy [22]. It was shown that FoxO1, 3, 4 triple knockout prevented muscle loss after fasting through inhibition of autophagy and protein ubiquitination. Inhibition of FoxOs decreases many important autophagy-related genes such as LC3, Gabarapl, Binp3, and P62/SQSTM1. A further study shows that FoxO1, 3, 4 are redundant in regulation of autophagy in skeletal muscle [22]. In addition, histone deacetylase 6 (HDAC6) can serve as the downstream target of FoxO3 in autophagy regulation [92]. During denervation, HDAC6 is upregulated during muscle atrophy and FoxO3 can directly bind to the HDAC6 promoter to regulate the expression of HDAC6, while knockdown of HDAC6 using shRNA reduces muscle atrophy (Fig. 4.3) [92]. The role of FoxOs and the insulin or insulin-like growth factor-1 (IGF-1) signaling pathway in autophagy is further confirmed in vivo. Muscle-specific knockout of the insulin receptor (M-IR$^{-/-}$), but not the IGF-1 receptor (M-IGF1R$^{-/-}$), displays a moderate reduction in muscle mass, while both IR and IGF-1R knockout mice (MIGIRKO) shows a marked reduction of muscle mass [93]. However, combined muscle-specific knockout FoxO1, 3, 4 in MIGIRKO mice reverses the increased autophagy and rescues the muscle mass loss [93]. These results further confirm the importance of insulin or IGF-1 mediated FoxOs' activity in autophagy in skeletal muscle. Other studies show that PGC1α and PGC1β can block the effect of FoxO3 or starvation on autophagy to prevent muscle atrophy, which indicates that exercise can increase the expression of PGC1α to inhibit FoxO3 activity to prevent muscle atrophy [94]. Interestingly, Wei et al. showed that mammalian sterile 20-like kinase 1 (MST1) can regulate FoxO3 activity through phosphorylation of FoxO3 at Ser207 [95]. MST1 kinase is upregulated in fast twitch skeletal muscles immediately after denervation. Activated MST1 increases FoxO3 phosphorylation at Ser207 and promotes FoxO3's nuclear translocation to induce autophagy. Notably, MST1-mediated phosphorylation of FoxO3 at Ser 207 can promote FoxO3 nuclear translocation, while pAkt-mediated phosphorylation of FoxO3 at Thr32, Ser253, and Ser315 can inhibit FoxO3 nuclear translocation. This difference may be due to the phosphorylation at different sites that impairs the interactions of FoxO3 with protein 14-3-3.

In addition to phosphorylation, acetylation also affects FoxO3 activity and localization. During fasting, the expression of Sirt1 is reduced dramatically in type II skeletal muscle, which leads to increased atrophy [96]. Sustained expression of

Sirt1 can prevent atrophy induced by fasting or denervation in the skeletal muscle of mice. Overexpression of Sirt1 can block the activity of FoxO1 and FoxO3 during atrophy and further prevent the induction of expression of autophagy genes. Sirt1-mediated FoxO1 and FoxO3 deacetylation inhibits their activity to induce autophagy during fasting or denervation in skeletal muscle. During denervation, FoxO1 is hyperacetylated, increasing cytosolic distribution of FoxO3 and its degradation via the proteasome system [97]. However, Hussain et al. show that in the diaphragm and limb muscles of humans, prolonged controlled mechanical ventilation (CMV) triggers autophagy, which is associated with increased gene expression of FoxO1, but not FoxO3 [98]. Recently, it was shown that calcitonin gene-related peptide (CGRP) can decrease FoxOs-mediated autophagy through protein kinase A (PKA)/CREB signaling in vitro and in vivo in skeletal muscle (Fig. 4.3) [99]. CGRP can elevate cAMP levels, stimulate PKA/CREB signaling, and increase FoxO1 phosphorylation to reduce the FoxO1 activity on autophagy-related gene expression in a concentration-dependent manner, while PKA inhibitors can abolish the effect of CGRP on FoxO1, 3, 4 and autophagy [99].

### *4.3.5 FoxO–Autophagy Axis in the Heart*

FoxO transcription factors were found to regulate autophagy in the heart under starvation or ischemia–reperfusion in mice (Fig. 4.5). In cultured rat neonatal cardiomyocytes, overexpression of either FoxO1 or FoxO3 reduced cell size and induced the autophagy pathway, showing increased activity of genes *LC3, Gabarapl*, and Atg12 (Fig. 4.5). These effects of overexpression of either FoxO1 or FoxO3 were similar to the condition of glucose deprivation [80]. Moreover, FoxO1 and FoxO3 directly bound to the promoters of *Gabarapl* and Atg12. Inhibition of FoxO1 activity by overexpression of dominate negative FoxO1 (Δ256) reversed the effect of starvation on cardiomyocyte size [80]. Under cellular stress, such as starvation or ischemia–reperfusion in vivo, autophagy was induced accompanied with increased FoxO1 and FoxO3 activity. Thus, FoxO1 and FoxO3 regulated autophagy and cell size in cardiomyocytes [80]. A further study showed that Sirt1 induced deacetylation of FoxO1, while the upregulation of Rab7 also played a role in starvation-induced autophagy in cardiomyocytes [100]. FoxO3 induced autophagy via AMPK signaling pathway under conditions of hypoxia in H9C2 cells [24]. Insulin suppressed autophagy-related genes such as LC3 and *Gabarapl* in cardiomyocytes through phosphorylation of Akt and downstream FoxO3, while acute insulin deficiency caused by streptozotocin (STZ) and increased autophagy genes LC3 and *Gabarapl* as well as the muscle-specific Ub-ligases atrogin-1 and MuRF1 [101]. A

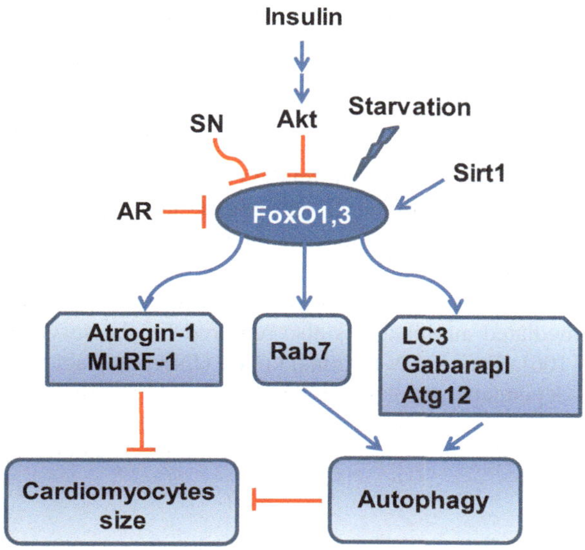


**Fig. 4.5** FoxOs regulate autophagy and cardiomyocytes size in the heart. During starvation, FoxOs are activated and induce autophagy through upregulating LC3, Gabarapl, Atg12, and Rab7. The upregulated autophagy by FoxOs can reduce cardiomyocytes size. FoxOs can also activate ubiquitin–proteasome system by increasing atrogin-1 and MuRF-1 to reduce cell size. Insulin signaling, SN, and AR can inhibit FoxO-mediated reduction of cardiomyocyte size. See the text in Sect. 4.3.5 for detailed discussion. *AR* adrenoceptor, *SN* sympathetic neuron

further study showed that cardiac sympathetic neuron (SN) ablation caused a reduction in cardiomyocyte size through FoxO-mediated atrogin-1 and MuRF1 but decreased stimulation of cardiomyocyte β2-adrenoceptor (β2-AR) (Fig. 4.5) [102]. Consistently, β2-AR agonist clenbuterol treatment prevented atrophy in denervated mice, while β2-AR knockout mice showed cardiac atrophy [102]. FoxO transcription factors played roles in oxidative stress resistance in cardiomyocytes. FoxO1 and FoxO3 and their target genes' activities were promoted by oxidative stress. Overexpression of FoxO1 and FoxO3 reduced reactive oxidative species (ROS) and cell death, while dominate-negative FoxO1 (Δ256) increased ROS and cell death significantly in cardiomyocytes. Cardiomyocyte-specific FoxO1/3 knockout mice subjected to myocardial infarction (MI) showed reduced cardiac function and increased ROS and cell death compared to control. Thus, FoxO1 and FoxO3 promote cardiomyocyte survival under oxidative stress [103]. A further study confirmed the role of FoxO1 in protecting cardiomyocytes from oxidative stress in cardiomyocyte H9C2 cells [104].

### 4.3.6 FoxO–Autophagy Axis in Other Tissues and Medical Conditions

#### 4.3.6.1 FoxOs Regulate Autophagy in Neurology

FoxO is shown to play a role in the elimination of neural stem cells (neuroblasts) in Drosophila [105]. Inhibition of FoxO and reaper family proapoptotic genes can increase the survival of neuroblasts and maintain neurogenesis in adult mushroom body, which is possibly through autophagy and apoptosis, since inhibition of autophay-related gene Atg1 and apoptosis can also promote neuroblast survival [105]. FoxO1-mediated autophagy is also required for the survival of neurons in mice or MEFs [106]. The cJun N-terminal kinase (JNK) may act as a negative regulator of FoxO1-dependent autophagy in neurons. Triple ablation of Jnk1, Jnk2, and Jnk3 in neurons can increase the nuclear localization of FoxO1 and its target gene Bnip3 to activate autophagy and increase the neuron life span [106].

#### 4.3.6.2 FoxOs Regulate Autophagy in Immunity

The role of DAF-16 (a homologous protein of FoxO) in pathogen resistance via autophagy was initially illustrated in C. elegans. Overexpression of DAF-16 increased autophagy in TJ356 animals (a strain of C. elegans) indicated by an increased number of GFP::LGG1 punctuate dots [107]. This DAF-16 overexpressed TJ356 showed resistance to *Salmonella* infection, while this pathogen resistance was blocked by the autophagy genes *bec-1* and *lgg-1* RNAi [107]. Thus, DAF-16 (FoxO) transcription factor induced autophagy to resistant pathogen infection. Recently, one study showed that FoxO1-mediated autophagy was required for natural killer (NK) cell maturation. Strong autophagy was found in immature NK (iNK) cells, while autophagy deficiency in NK-specific Atg5 knockout mice (*Atg5*$^{flox/flox}$, *NKp46-Cre*) resulted in damaged mitochondria and an increase in ROS, as well as cell death [108]. Interestingly, phosphorylated FoxO1 in the cytoplasm of iNK interacted with Atg7 to induce autophagy, which is independent of the transcriptional activity of FoxO1 [108]. Of note, another group showed that FoxO1 regulated NK maturation negatively [109]. Further studies are needed to confirm the role of FoxO1 in NK maturation.

#### 4.3.6.3 FoxOs Regulate Autophagy in Aging

It was shown that dFoxO regulates autophagy to affect longevity in Drosophila [110]. While the characterization of muscle aging in Drosophila is ongoing, accumulation of protein aggregates and overexpression of dFoxO in the muscle can delay the accumulation of protein aggregates, at least partly through autophagy [110]. Both dFoxO and its target 4E-BP can delay aging-related muscle function decay and expand life span in Drosophila. Other studies show that dFoxO may activate autophagy through inhibition of the Activin signaling pathway to expand the life span in Drosophila [111]. DFoxO1 can bind and repress dawdle, an Activin ligand, to prevent the inhibition of the Smad binding element on autophagy-related gene Atg8a, leading to an increased life span [111]. During human joint aging from 23 to 90, the expression of FoxO1 and FoxO3 are markedly decreased in the superficial zone of cartilage. In the cartilage during osteoarthritis (OA), the activity of FoxO is highly inhibited with increased phosphorylation of FoxO1 and increased cytosolic localization [112]. Since FoxOs can activate autophagy in many tissues, it is possible that FoxOs-mediated autophagy is decreased during the aging of knee joints.

#### 4.3.6.4 Tumorigenesis and Cancer Cells

FoxOs can regulate autophagy to inhibit colorectal cancer growth [113]. Previous studies show that the activity of p38α is required for proliferation and survival of colorectal cancer cell (CRC). P38α can increase the expression of hypoxia-inducible factor 1α (HIF1α) and its target glycolytic rate-limiting genes such as GLUT1, Hexokinase (HK) 1/2, Pyruvate kinase isozymes M2 (PKM2) and Lactate dehydrogenase A (LDHA). Blockage of p38α using a p38α inhibitor SB202190 can decrease the expression of HIF1α and its target genes, while activation of FoxO3 and its target autophagy-related genes, such as MAP1LC3, GABARAPL1, ATG12, BNIP3, and BNIP3L induces autophagy and inhibits colorectal cancer growth. Similar results are also shown in ovarian cancer cells [114]. Histone deacetylase inhibitors (HDACIs) can induce autophagy through increasing the expression of FoxO1 and its transcriptional activity in HCT116 colon cancer cells [115]. Knockdown of FoxO1 using siRNA can block HDACI-induced autophagy. These data show that FoxO1 and FoxO3 can induce autophagy in different cancer cells. Interestingly, it is shown that cytosolic FoxO1 is required for its induction of FoxO1 and tumor suppressor activity which is independent of its transcriptional activity in different cancer cell lines (such as HCT116, Hela, and H1299 cells) [20]. Under oxidative stress or serum starvation, cytosolic FoxO1 can dissociate with Sirt 2 which leads to the acetylation of FoxO1. The acetylated FoxO1 binds to Atg7 to induce autophagy [20]. Furthermore, FoxO3 can induce autophagy in a FoxO1-dependent way at least in the HEK293T human embryonic kidney cell line and mouse embryonic fibroblast (MEF) cell lines [19]. Controversially, other study shows that FoxO3 can negatively regulate autophagy in PC3 (prostate cancer) cells, HCT116 (colon cancer), and MDA-MB-231 (breast cancer) cell lines, as FoxO3 inhibits the expression of FoxO1 [116]. These controversial data from different research groups indicate that the roles of FoxO1 and FoxO3 in autophagy of cancer cells are complicated and may be due to their cell type-dependent expression pattern or specific signaling pathways.

## 4.4 Conclusions

Metabolic health is critical for tissue homeostasis. As transcription factors, the FoxO proteins control the expression of an array of genes, including the genes regulating metabolic pathways, autophagy, cell cycle arrest, and differentiation. The FoxO–autophagy axis has been identified across tissues such as liver, skeletal muscle, heart, and adipose tissue (or adipocytes). FoxO1 and FoxO3 can directly regulate the expression of Atg12, Atg14, Binp3, Gabarapl, LC3, and P62/SQSTM1, suggesting that FoxOs regulate autophagy in multiple stages of the process. In addition, FoxO1 induces autophagy during adipocyte differentiation via Tfeb, the transcription factor that governs expression of autophagosome and lysosome genes. Blockage of the FoxO1-autophagy axis suppresses adipocyte differentiation, thereby inhibiting adipose tissue development. Suppression of FoxO or autophagy in the skeletal muscle potently prevents muscle atrophy, further corroborating the important role of FoxO–autophagy axis in tissue homeostasis. Given that FoxO transcription factors are activated by insulin resistance, one of the hallmarks of metabolic diseases [8], targeting the FoxO–autophagy axis may provide new options to preserve metabolic health and tissue homeostasis.

**Acknowledgments** This work was supported by USDA National Institute of Food and Agriculture Hatch Project 1007334 (ZC) and American Heart Association Grant 18TPA34230082 (ZC) . The authors declare no potential conflict of interest.

## References

1. Jacobs FM, van der Heide LP, Wijchers PJ, Burbach JP, Hoekman MF, Smidt MP. FoxO6, a novel member of the FoxO class of transcription factors with distinct shuttling dynamics. J Biol Chem. 2003;278:35959–67.
2. Kaestner KH, Knochel W, Martinez DE. Unified nomenclature for the winged helix/forkhead transcription factors. Genes Dev. 2000;14:142–6.
3. Kenyon C, Chang J, Gensch E, Rudner A, Tabtiang R. A C. elegans mutant that lives twice as long as wild type. Nature. 1993;366:461–4.
4. Lin K, Dorman JB, Rodan A, Kenyon C. daf-16: An HNF-3/forkhead family member that can function to double the life-span of Caenorhabditis elegans. Science. 1997;278:1319–22.
5. Ogg S, Paradis S, Gottlieb S, et al. The Fork head transcription factor DAF-16 transduces insulin-like metabolic and longevity signals in C. elegans. Nature. 1997;389:994–9.
6. Furuyama T, Nakazawa T, Nakano I, Mori N. Identification of the differential distribution patterns of mRNAs and consensus binding sequences for mouse DAF-16 homologues. Biochem J. 2000;349:629–34.
7. Obsil T, Obsilova V. Structure/function relationships underlying regulation of FOXO transcription factors. Oncogene. 2008;27:2263–75.
8. Cheng Z, White MF. Targeting Forkhead box O1 from the concept to metabolic diseases: lessons from mouse models. Antioxid Redox Signal. 2011;14:649–61.
9. van der Vos KE, Coffer PJ. The extending network of FOXO transcriptional target genes. Antioxid Redox Signal. 2011;14:579–92.

10. Matsuzaki H, Daitoku H, Hatta M, Tanaka K, Fukamizu A. Insulin-induced phosphorylation of FKHR (Foxo1) targets to proteasomal degradation. Proc Natl Acad Sci U S A. 2003;100:11285–90.
11. Huang H, Regan KM, Wang F, et al. Skp2 inhibits FOXO1 in tumor suppression through ubiquitin-mediated degradation. Proc Natl Acad Sci U S A. 2005;102:1649–54.
12. Brunet A, Sweeney LB, Sturgill JF, et al. Stress-dependent regulation of FOXO transcription factors by the SIRT1 deacetylase. Science. 2004;303:2011–5.
13. Furuyama T, Kitayama K, Yamashita H, Mori N. Forkhead transcription factor FOXO1 (FKHR)-dependent induction of PDK4 gene expression in skeletal muscle during energy deprivation. Biochem J. 2003;375:365–71.
14. Nakae J, Kitamura T, Kitamura Y, Biggs WH 3rd, Arden KC, Accili D. The forkhead transcription factor Foxo1 regulates adipocyte differentiation. Dev Cell. 2003;4:119–29.
15. Zhang W, Patil S, Chauhan B, et al. FoxO1 regulates multiple metabolic pathways in the liver: effects on gluconeogenic, glycolytic, and lipogenic gene expression. J Biol Chem. 2006;281:10105–17.
16. Kim DH, Zhang T, Lee S, et al. FoxO6 integrates insulin signaling with MTP for regulating VLDL production in the liver. Endocrinology. 2014;155:1255–67.
17. Calabuig-Navarro V, Yamauchi J, Lee S, et al. Forkhead box O6 (FoxO6) depletion attenuates hepatic gluconeogenesis and protects against fat-induced glucose disorder in mice. J Biol Chem. 2015;290:15581–94.
18. Lee S, Dong HH. FoxO integration of insulin signaling with glucose and lipid metabolism. J Endocrinol. 2017;233:R67–79.
19. Zhou J, Liao W, Yang J, et al. FOXO3 induces FOXO1-dependent autophagy by activating the AKT1 signaling pathway. Autophagy. 2012;8:1712–23.
20. Zhao Y, Yang J, Liao W, et al. Cytosolic FoxO1 is essential for the induction of autophagy and tumour suppressor activity. Nat Cell Biol. 2010;12:665–75.
21. Zhao J, Brault JJ, Schild A, et al. FoxO3 coordinately activates protein degradation by the autophagic/lysosomal and proteasomal pathways in atrophying muscle cells. Cell Metab. 2007;6:472–83.
22. Milan G, Romanello V, Pescatore F, et al. Regulation of autophagy and the ubiquitin-proteasome system by the FoxO transcriptional network during muscle atrophy. Nat Commun. 2015;6:6670.
23. Mammucari C, Milan G, Romanello V, et al. FoxO3 controls autophagy in skeletal muscle in vivo. Cell Metab. 2007;6:458–71.
24. Chi Y, Shi C, Zhao Y, Guo C. Forkhead box O (FOXO) 3 modulates hypoxia-induced autophagy through AMPK signalling pathway in cardiomyocytes. Biosci Rep. 2016;36:e00345.
25. Liu L, Tao Z, Zheng LD, et al. FoxO1 interacts with transcription factor EB and differentially regulates mitochondrial uncoupling proteins via autophagy in adipocytes. Cell Death Dis. 2016;2:16066.
26. Liu L, Zheng LD, Zou P, et al. FoxO1 antagonist suppresses autophagy and lipid droplet growth in adipocytes. Cell Cycle. 2016;15:2033–41.
27. Lopategi A, Lopez-Vicario C, Alcaraz-Quiles J, et al. Role of bioactive lipid mediators in obese adipose tissue inflammation and endocrine dysfunction. Mol Cell Endocrinol. 2016;419:44–59.
28. Lanthier N, Leclercq IA. Adipose tissues as endocrine target organs. Best Pract Res Clin Gastroenterol. 2014;28:545–58.
29. Dowell P, Otto TC, Adi S, Lane MD. Convergence of peroxisome proliferator-activated receptor gamma and Foxo1 signaling pathways. J Biol Chem. 2003;278:45485–91.
30. Fan W, Imamura T, Sonoda N, et al. FOXO1 transrepresses peroxisome proliferator-activated receptor gamma transactivation, coordinating an insulin-induced feed-forward response in adipocytes. J Biol Chem. 2009;284:12188–97.

31. Armoni M, Harel C, Karni S, et al. FOXO1 represses peroxisome proliferator-activated receptor-gamma1 and -gamma2 gene promoters in primary adipocytes. A novel paradigm to increase insulin sensitivity. J Biol Chem. 2006;281:19881–91.
32. Nakae J, Cao Y, Oki M, et al. Forkhead transcription factor FoxO1 in adipose tissue regulates energy storage and expenditure. Diabetes. 2008;57:563–76.
33. Jing E, Gesta S, Kahn CR. SIRT2 regulates adipocyte differentiation through FoxO1 acetylation/deacetylation. Cell Metab. 2007;6:105–14.
34. Bai L, Pang WJ, Yang YJ, Yang GS. Modulation of Sirt1 by resveratrol and nicotinamide alters proliferation and differentiation of pig preadipocytes. Mol Cell Biochem. 2008;307:129–40.
35. Higuchi M, Dusting GJ, Peshavariya H, et al. Differentiation of human adipose-derived stem cells into fat involves reactive oxygen species and Forkhead box O1 mediated upregulation of antioxidant enzymes. Stem Cells Dev. 2013;22:878–88.
36. Kim H, Hiraishi A, Tsuchiya K, Sakamoto K. (-) Epigallocatechin gallate suppresses the differentiation of 3T3-L1 preadipocytes through transcription factors FoxO1 and SREBP1c. Cytotechnology. 2010;62:245–55.
37. Munekata K, Sakamoto K. Forkhead transcription factor Foxo1 is essential for adipocyte differentiation. In Vitro Cell Dev Biol Anim. 2009;45:642–51.
38. Zou P, Liu L, Zheng L, et al. Targeting FoxO1 with AS1842856 suppresses adipogenesis. Cell Cycle. 2014;13:3759–67.
39. Gross DN, van den Heuvel AP, Birnbaum MJ. The role of FoxO in the regulation of metabolism. Oncogene. 2008;27:2320–36.
40. Samuel VT, Choi CS, Phillips TG, et al. Targeting foxo1 in mice using antisense oligonucleotide improves hepatic and peripheral insulin action. Diabetes. 2006;55:2042–50.
41. Titchenell PM, Chu Q, Monks BR, Birnbaum MJ. Hepatic insulin signalling is dispensable for suppression of glucose output by insulin in vivo. Nat Commun. 2015;6:7078.
42. Matsumoto M, Han S, Kitamura T, Accili D. Dual role of transcription factor FoxO1 in controlling hepatic insulin sensitivity and lipid metabolism. J Clin Invest. 2006;116:2464–72.
43. Kamagate A, Qu S, Perdomo G, et al. FoxO1 mediates insulin-dependent regulation of hepatic VLDL production in mice. J Clin Invest. 2008;118:2347–64.
44. Altomonte J, Cong L, Harbaran S, et al. Foxo1 mediates insulin action on apoC-III and triglyceride metabolism. J Clin Invest. 2004;114:1493–503.
45. Qu S, Su D, Altomonte J, et al. PPAR{alpha} mediates the hypolipidemic action of fibrates by antagonizing FoxO1. Am J Physiol Endocrinol Metab. 2007;292:E421–34.
46. Zhang K, Li L, Qi Y, et al. Hepatic suppression of Foxo1 and Foxo3 causes hypoglycemia and hyperlipidemia in mice. Endocrinology. 2012;153:631–46.
47. Tao R, Xiong X, DePinho RA, Deng CX, Dong XC. Hepatic SREBP-2 and cholesterol biosynthesis are regulated by FoxO3 and Sirt6. J Lipid Res. 2013;54:2745–53.
48. de Lange P, Moreno M, Silvestri E, Lombardi A, Goglia F, Lanni A. Fuel economy in food-deprived skeletal muscle: signaling pathways and regulatory mechanisms. FASEB J. 2007;21:3431–41.
49. Sugita S, Kamei Y, Akaike F, et al. Increased systemic glucose tolerance with increased muscle glucose uptake in transgenic mice overexpressing RXRgamma in skeletal muscle. PLoS One. 2011;6:e20467.
50. Kamei Y, Mizukami J, Miura S, et al. A forkhead transcription factor FKHR up-regulates lipoprotein lipase expression in skeletal muscle. FEBS Lett. 2003;536:232–6.
51. Bastie CC, Nahle Z, McLoughlin T, et al. FoxO1 stimulates fatty acid uptake and oxidation in muscle cells through CD36-dependent and -independent mechanisms. J Biol Chem. 2005;280:14222–9.
52. Kamei Y, Miura S, Suganami T, et al. Regulation of SREBP1c gene expression in skeletal muscle: role of retinoid X receptor/liver X receptor and forkhead-O1 transcription factor. Endocrinology. 2008;149:2293–305.

53. Nakashima K, Yakabe Y. AMPK activation stimulates myofibrillar protein degradation and expression of atrophy-related ubiquitin ligases by increasing FOXO transcription factors in C2C12 myotubes. Biosci Biotechnol Biochem. 2007;71:1650–6.
54. Manfredi LH, Zanon NM, Garofalo MA, Navegantes LC. Kettelhut IC (2013) Effect of short-term cold exposure on skeletal muscle protein breakdown in rats. J Appl Physiol. 1985;115:1496–505.
55. Kamei Y, Miura S, Suzuki M, et al. Skeletal muscle FOXO1 (FKHR) transgenic mice have less skeletal muscle mass, down-regulated Type I (slow twitch/red muscle) fiber genes, and impaired glycemic control. J Biol Chem. 2004;279:41114–23.
56. Murtaugh LC. Pancreas and beta-cell development: from the actual to the possible. Development. 2007;134:427–38.
57. Straub SG, Sharp GW. Glucose-stimulated signaling pathways in biphasic insulin secretion. Diabetes Metab Res Rev. 2002;18:451–63.
58. Kitamura T, Nakae J, Kitamura Y, et al. The forkhead transcription factor Foxo1 links insulin signaling to Pdx1 regulation of pancreatic beta cell growth. J Clin Invest. 2002;110:1839–47.
59. Kobayashi M, Kikuchi O, Sasaki T, et al. FoxO1 as a double-edged sword in the pancreas: analysis of pancreas- and beta-cell-specific FoxO1 knockout mice. Am J Physiol Endocrinol Metab. 2012;302:E603–13.
60. Gupta D, Leahy AA, Monga N, Peshavaria M, Jetton TL, Leahy JL. Peroxisome proliferator-activated receptor gamma (PPARgamma) and its target genes are downstream effectors of FoxO1 protein in islet beta-cells: mechanism of beta-cell compensation and failure. J Biol Chem. 2013;288:25440–9.
61. Liu Y, Wang X, Wu H, et al. Glycine enhances muscle protein mass associated with maintaining Akt-mTOR-FOXO1 signaling and suppressing TLR4 and NOD2 signaling in piglets challenged with LPS. Am J Physiol Regul Integr Comp Physiol. 2016;311:R365–73.
62. Kitamura YI, Kitamura T, Kruse JP, et al. FoxO1 protects against pancreatic beta cell failure through NeuroD and MafA induction. Cell Metab. 2005;2:153–63.
63. Kawamori D, Kaneto H, Nakatani Y, et al. The forkhead transcription factor Foxo1 bridges the JNK pathway and the transcription factor PDX-1 through its intracellular translocation. J Biol Chem. 2006;281:1091–8.
64. Tanida I. Autophagosome formation and molecular mechanism of autophagy. Antioxid Redox Signal. 2011;14:2201–14.
65. Mizushima N, Yoshimori T, Ohsumi Y. The role of Atg proteins in autophagosome formation. Annu Rev Cell Dev Biol. 2011;27:107–32.
66. Itakura E, Kishi C, Inoue K, Mizushima N. Beclin 1 forms two distinct phosphatidylinositol 3-kinase complexes with mammalian Atg14 and UVRAG. Mol Biol Cell. 2008;19:5360–72.
67. Matsunaga K, Saitoh T, Tabata K, et al. Two Beclin 1-binding proteins, Atg14L and Rubicon, reciprocally regulate autophagy at different stages. Nat Cell Biol. 2009;11:385–96.
68. Obara K, Noda T, Niimi K, Ohsumi Y. Transport of phosphatidylinositol 3-phosphate into the vacuole via autophagic membranes in Saccharomyces cerevisiae. Genes Cells. 2008;13:537–47.
69. Sun Q, Fan W, Chen K, Ding X, Chen S, Zhong Q. Identification of Barkor as a mammalian autophagy-specific factor for Beclin 1 and class III phosphatidylinositol 3-kinase. Proc Natl Acad Sci U S A. 2008;105:19211–6.
70. Sun Q, Fan W, Zhong Q. Regulation of Beclin 1 in autophagy. Autophagy. 2009;5:713–6.
71. Itakura E, Mizushima N. Characterization of autophagosome formation site by a hierarchical analysis of mammalian Atg proteins. Autophagy. 2010;6:764–76.
72. Kim J, Kundu M, Viollet B, Guan KL. AMPK and mTOR regulate autophagy through direct phosphorylation of Ulk1. Nat Cell Biol. 2011;13:132–41.
73. Egan DF, Shackelford DB, Mihaylova MM, et al. Phosphorylation of ULK1 (hATG1) by AMP-activated protein kinase connects energy sensing to mitophagy. Science. 2011;331:456–61.
74. Mizushima N, Yamamoto A, Hatano M, et al. Dissection of autophagosome formation using Apg5-deficient mouse embryonic stem cells. J Cell Biol. 2001;152:657–68.

75. Hanada T, Noda NN, Satomi Y, et al. The Atg12-Atg5 conjugate has a novel E3-like activity for protein lipidation in autophagy. J Biol Chem. 2007;282:37298–302.
76. Bjorkoy G, Lamark T, Brech A, et al. p62/SQSTM1 forms protein aggregates degraded by autophagy and has a protective effect on huntingtin-induced cell death. J Cell Biol. 2005;171:603–14.
77. Pankiv S, Clausen TH, Lamark T, et al. p62/SQSTM1 binds directly to Atg8/LC3 to facilitate degradation of ubiquitinated protein aggregates by autophagy. J Biol Chem. 2007;282:24131–45.
78. Rogov V, Dotsch V, Johansen T, Kirkin V. Interactions between autophagy receptors and ubiquitin-like proteins form the molecular basis for selective autophagy. Mol Cell. 2014;53:167–78.
79. Xiong X, Tao R, DePinho RA, Dong XC. The autophagy-related gene 14 (Atg14) is regulated by forkhead box O transcription factors and circadian rhythms and plays a critical role in hepatic autophagy and lipid metabolism. J Biol Chem. 2012;287:39107–14.
80. Sengupta A, Molkentin JD, Yutzey KE. FoxO transcription factors promote autophagy in cardiomyocytes. J Biol Chem. 2009;284:28319–31.
81. Settembre C, Di Malta C, Polito VA, et al. TFEB links autophagy to lysosomal biogenesis. Science. 2011;332:1429–33.
82. Settembre C, De Cegli R, Mansueto G, et al. TFEB controls cellular lipid metabolism through a starvation-induced autoregulatory loop. Nat Cell Biol. 2013;15:647–58.
83. Singh R, Xiang Y, Wang Y, et al. Autophagy regulates adipose mass and differentiation in mice. J Clin Invest. 2009;119:3329–39.
84. Zhang Y, Goldman S, Baerga R, Zhao Y, Komatsu M, Jin S. Adipose-specific deletion of autophagy-related gene 7 (atg7) in mice reveals a role in adipogenesis. Proc Natl Acad Sci U S A. 2009;106:19860–5.
85. Zhang C, He Y, Okutsu M, et al. Autophagy is involved in adipogenic differentiation by repressesing proteasome-dependent PPARgamma2 degradation. Am J Physiol Endocrinol Metab. 2013;305:E530–9.
86. Puri V, Ranjit S, Konda S, et al. Cidea is associated with lipid droplets and insulin sensitivity in humans. Proc Natl Acad Sci U S A. 2008;105:7833–8.
87. Wang H, Liu L, Lin JZ, Aprahamian TR, Farmer SR. Browning of white adipose tissue with roscovitine induces a distinct population of UCP1+ adipocytes. Cell Metab. 2016;24:835–47.
88. Taylor D, Gottlieb RA. Parkin-mediated mitophagy is downregulated in browning of white adipose tissue. Obesity (Silver Spring). 2017;25:704–12.
89. Kalinovich AV, de Jong JM, Cannon B, Nedergaard J. UCP1 in adipose tissues: two steps to full browning. Biochimie. 2017;134:127–37.
90. Xu L, Kanasaki M, He J, et al. Ketogenic essential amino acids replacement diet ameliorated hepatosteatosis with altering autophagy-associated molecules. Biochim Biophys Acta. 2013;1832:1605–12.
91. Song YM, Lee YH, Kim JW, et al. Metformin alleviates hepatosteatosis by restoring SIRT1-mediated autophagy induction via an AMP-activated protein kinase-independent pathway. Autophagy. 2015;11:46–59.
92. Ratti F, Ramond F, Moncollin V, et al. Histone deacetylase 6 is a FoxO transcription factor-dependent effector in skeletal muscle atrophy. J Biol Chem. 2015;290:4215–24.
93. O'Neill BT, Lee KY, Klaus K, et al. Insulin and IGF-1 receptors regulate FoxO-mediated signaling in muscle proteostasis. J Clin Invest. 2016;126:3433–46.
94. Brault JJ, Jespersen JG, Goldberg AL. Peroxisome proliferator-activated receptor gamma coactivator 1alpha or 1beta overexpression inhibits muscle protein degradation, induction of ubiquitin ligases, and disuse atrophy. J Biol Chem. 2010;285:19460–71.
95. Wei B, Dui W, Liu D, Xing Y, Yuan Z, Ji G. MST1, a key player, in enhancing fast skeletal muscle atrophy. BMC Biol. 2013;11:12.

96. Lee D, Goldberg AL. SIRT1 protein, by blocking the activities of transcription factors FoxO1 and FoxO3, inhibits muscle atrophy and promotes muscle growth. J Biol Chem. 2013;288:30515–26.
97. Bertaggia E, Coletto L, Sandri M. Posttranslational modifications control FoxO3 activity during denervation. Am J Physiol Cell Physiol. 2012;302:C587–96.
98. Hussain SN, Mofarrahi M, Sigala I, et al. Mechanical ventilation-induced diaphragm disuse in humans triggers autophagy. Am J Respir Crit Care Med. 2010;182:1377–86.
99. Machado J, Manfredi LH, Silveira WA, et al. Calcitonin gene-related peptide inhibits autophagic-lysosomal proteolysis through cAMP/PKA signaling in rat skeletal muscles. Int J Biochem Cell Biol. 2016;72:40–50.
100. Hariharan N, Maejima Y, Nakae J, Paik J, Depinho RA, Sadoshima J. Deacetylation of FoxO by Sirt1 plays an essential role in mediating starvation-induced autophagy in cardiac myocytes. Circ Res. 2010;107:1470–82.
101. Paula-Gomes S, Goncalves DA, Baviera AM, Zanon NM, Navegantes LC, Kettelhut IC. Insulin suppresses atrophy- and autophagy-related genes in heart tissue and cardiomyocytes through AKT/FOXO signaling. Horm Metab Res. 2013;45:849–55.
102. Zaglia T, Milan G, Franzoso M, et al. Cardiac sympathetic neurons provide trophic signal to the heart via beta2-adrenoceptor-dependent regulation of proteolysis. Cardiovasc Res. 2013;97:240–50.
103. Sengupta A, Molkentin JD, Paik JH, DePinho RA, Yutzey KE. FoxO transcription factors promote cardiomyocyte survival upon induction of oxidative stress. J Biol Chem. 2011;286:7468–78.
104. Ning Y, Li Z, Qiu Z. FOXO1 silence aggravates oxidative stress-promoted apoptosis in cardiomyocytes by reducing autophagy. J Toxicol Sci. 2015;40:637–45.
105. Siegrist SE, Haque NS, Chen CH, Hay BA, Hariharan IK. Inactivation of both Foxo and reaper promotes long-term adult neurogenesis in Drosophila. Curr Biol. 2010;20:643–8.
106. Xu P, Das M, Reilly J, Davis RJ. JNK regulates FoxO-dependent autophagy in neurons. Genes Dev. 2011;25:310–22.
107. Jia K, Thomas C, Akbar M, et al. Autophagy genes protect against Salmonella typhimurium infection and mediate insulin signaling-regulated pathogen resistance. Proc Natl Acad Sci U S A. 2009;106:14564–9.
108. Wang S, Xia P, Huang G, et al. FoxO1-mediated autophagy is required for NK cell development and innate immunity. Nat Commun. 2016;7:11023.
109. Deng Y, Kerdiles Y, Chu J, et al. Transcription factor Foxo1 is a negative regulator of natural killer cell maturation and function. Immunity. 2015;42:457–70.
110. Demontis F, Perrimon N. FOXO/4E-BP signaling in Drosophila muscles regulates organism-wide proteostasis during aging. Cell. 2010;143:813–25.
111. Bai H, Kang P, Hernandez AM, Tatar M. Activin signaling targeted by insulin/dFOXO regulates aging and muscle proteostasis in Drosophila. PLoS Genet. 2013;9:e1003941.
112. Akasaki Y, Hasegawa A, Saito M, Asahara H, Iwamoto Y, Lotz MK. Dysregulated FOXO transcription factors in articular cartilage in aging and osteoarthritis. Osteoarthritis Cartilage. 2014;22:162–70.
113. Chiacchiera F, Matrone A, Ferrari E, et al. p38alpha blockade inhibits colorectal cancer growth in vivo by inducing a switch from HIF1alpha- to FoxO-dependent transcription. Cell Death Differ. 2009;16:1203–14.
114. Matrone A, Grossi V, Chiacchiera F, et al. p38alpha is required for ovarian cancer cell metabolism and survival. Int J Gynecol Cancer. 2010;20:203–11.
115. Zhang J, Ng S, Wang J, et al. Histone deacetylase inhibitors induce autophagy through FOXO1-dependent pathways. Autophagy. 2015;11:629–42.
116. Zhu WL, Tong H, Teh JT, Wang M. Forkhead box protein O3 transcription factor negatively regulates autophagy in human cancer cells by inhibiting forkhead box protein O1 expression and cytosolic accumulation. PLoS One. 2014;9:e115087.

# Chapter 5
# The Role of mTOR in Osteoclasts

**Ralph A. Zirngibl and Irina Voronov**

**Abstract** Evolutionary conserved kinase mechanistic target of rapamycin (mTOR) is the signaling hub for cellular responses to nutrients, cytokines, growth hormones, and environmental stresses in all eukaryotic cells. Increased mTOR activity has been demonstrated in numerous diseases, such as cancer and autoimmune diseases. Due to its prominent role, mTOR inhibitors are being used and tested to treat a wide variety of conditions. Recent evidence suggests that regulation of mTOR activity and function is not universal and varies between the cells. Here we summarize the latest research on the role and regulation of mTOR in osteoclasts, the unique multinucleated bone-resorbing cells, focusing on the role of mTOR as part of the mTORC1 complex. Collectively, the results suggest that mTORC1 activity plays a double role in osteoclastogenesis: at the earlier stage, it is necessary for proliferation of the precursors, and, at the later stage, it is indispensable for cytoskeletal reorganization involved in the process of bone resorption. We also present evidence that in osteoclasts, mTOR protein levels and activity are regulated differently compared to other primary cells and cell lines. Due to this prominent role of mTOR in osteoclast formation and function, mTOR inhibitors could be used to treat numerous diseases that involve overactive osteoclasts, such as osteoporosis, inflammatory arthritis, Paget's disease, and cancer-related osteolysis.

## 5.1 Osteoclasts

The skeleton is constantly being remodeled. New bone is deposited by osteoblasts, the bone forming cells, while old or damaged bone is removed by osteoclasts, the bone resorbing cells. These cycles of bone formation and resorption are tightly controlled, with both osteoblast and osteoclasts secreting molecules regulating each other's activity (reviewed in [1, 2]). Osteoclasts are multinucleated cells of hematopoietic origin formed by fusion of mononuclear precursors (Fig. 5.1) [1, 3]. This

R. A. Zirngibl · I. Voronov (✉)
Faculty of Dentistry, University of Toronto, Toronto, ON, Canada
e-mail: irina.voronov@utoronto.ca



K. Turksen (ed.), *Autophagy in Health and Disease*, Stem Cell Biology and Regenerative Medicine, https://doi.org/10.1007/978-3-319-98146-8_5

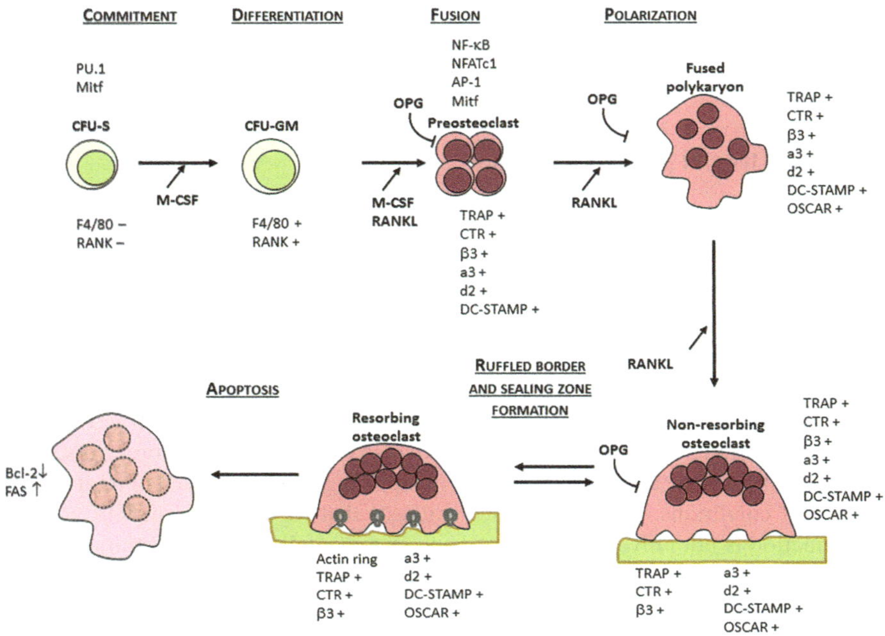


**Fig. 5.1** Osteoclastogenesis (adapted from Boyle et al. [69]). In the presence of M-CSF and RANKL, osteoclast precursors undergo differentiation and fusion. Transcription factors are listed above the cells; key functional proteins are listed below the cells. To regulate osteoclast formation and function, osteoblasts and stromal cells secrete osteoprotegerin (OPG), a decoy receptor for RANKL

precursor differentiation and fusion is initiated by two factors secreted by osteoblasts, osteocytes and stromal cells: macrophage colony stimulating factor (M-CSF) and receptor activator of nuclear factor κB ligand (RANKL). These two molecules are absolutely necessary for osteoclast differentiation, fusion, activity, and survival. Lack of either RANKL, its receptor RANK, M-CSF or M-CSF receptor CSF-1R leads to an osteoclastogenesis defect and severe osteopetrosis [4–6].

To resorb bone, mature multinucleated osteoclasts attach to the bone surface and form a tight sealing zone. This sealing zone is defined by a dense cytoskeletal actin ring structure, composed primarily of F-actin. Within this sealing zone, osteoclasts form a convoluted plasma membrane called a "ruffled border," a dynamic structure formed by continuous fusion of lysosomes and secretory vesicles delivering proteolytic enzymes for bone resorption, as well as continuous fission of transcytosing vesicles moving the degraded matrix away from the resorption site to the opposite (basolateral) side of the cell [1, 7]. The ruffled border is enriched with proton pumping vacuolar $H^+$-ATPases (V-ATPases) and chloride proton exchangers (ClC7), the protein complexes responsible for creating an acidic environment necessary to dissolve the mineral component of bone and to allow degradation of the bone matrix proteins [1].

M-CSF is responsible for osteoclast precursor proliferation, precursor commitment, cytoskeletal organization, and survival. M-CSF binding to its tyrosine kinase receptor CSF-1R (also known as c-Fms or M-CSF receptor) activates the phosphatidyl-inositol 3-kinase (PI3K)/AKT pathway, similar to other receptor tyrosine kinases. One of the downstream targets of PI3K/AKT pathway is a serine/threonine kinase mechanistic target of rapamycin (mTOR), the main subject of this chapter (its activation and function will be discussed later in greater detail). RANKL binding to RANK initiates a number of signaling cascades and activates several transcription factors, such as nuclear factor κB (NF-κB), activator protein 1 (AP-1) and nuclear factor of activated T cells c1 (NFATc1) [1, 3, 8, 9]. These transcription factors control transcription of osteoclast-specific genes that play a role in osteoclast function (e.g., tartrate-resistant acid phosphatase (TRAP), cathepsin K (CtsK), calcitonin receptor (CTR), osteoclast-enriched V-ATPase subunits *a*3 and d2); attachment (e.g., integrin $\alpha_v\beta_3$); or fusion (e.g., dendritic cell-specific transmembrane protein (DC-STAMP) and osteoclast-associated receptor (OSCAR)) as outlined in Fig. 5.1 [1, 10, 11].

Osteoclastogenesis is usually described as a multistage process which includes proliferation of the precursors, commitment, fusion of the committed osteoclast precursors, polarization on the bone surface, formation of the sealing zone/ruffled border, and apoptosis (although the latest intravital imaging shows fission of osteoclasts at the end of the bone resorbing cycle [12]; therefore, it is possible that apoptosis is not the ultimate last stage of the osteoclast life cycle in vivo). Each stage of osteoclastogenesis is defined by the expression of key proteins—transcription factors and other proteins involved in osteoclast differentiation, fusion, and function. For example, PU.1 is the earliest hematopoietic transcription factor expressed by the osteoclast precursors, and loss of PU.1 results in the complete absence of osteoclasts and myeloid precursors [8]. To elucidate precise molecular mechanisms activated during different stages of osteoclast differentiation, two conditional gene targeting mouse models are widely used. The lysozyme M (*LysM*)-Cre mouse model targets osteoclast precursors, since LysM is expressed mainly by the cells of the myeloid lineage, the cells that include osteoclast progenitors, monocytes, macrophages, and dendritic cells [13]. Meanwhile, the *Ctsk*-Cre mouse model targets later stages of osteoclastogenesis, since CtsK is expressed primarily by mature osteoclasts and not by the precursors [14].

As mentioned earlier, M-CSF signaling through PI3K/AKT activates mTOR, an evolutionary conserved kinase responsible for cellular responses to growth factors, nutrient availability and other extracellular cues [15]. It regulates protein and lipid synthesis, lysosomal and mitochondrial biogenesis, just to name a few, in all eukaryotic cells. The purpose of this chapter is to summarize the latest research on the role of mTOR in osteoclast differentiation and function. But, first, we will briefly describe the major players involved in mTOR signaling.

## 5.2 Overview of mTOR Signaling

mTOR belongs to the PI3K-related family of kinases. As the name implies, TOR (target of rapamycin) was identified in yeast genetic screens as the protein target of rapamycin, a macrolide with an antifungal and immunosuppressant activity [16]. In mammalian cells, mTOR exists as part of two multiprotein complexes, complex 1 (mTORC1) and complex 2 (mTORC2) (Fig. 5.2). mTORC1 consists of mTOR, regulatory-associated protein of mTOR (Raptor), DEP-domain containing mTOR-interaction protein (DEPTOR), proline-rich AKT substrate 40 kDa (PRAS40), and mammalian lethal with Sec13 protein 8 (mLST8). The functions of the mTORC1 components are well known: Raptor assists with substrate recognition and recruitment [17, 18], mLST8 is a positive regulator of mTOR activity [19], while PRAS40 and DEPTOR are the negative regulators of mTOR activity [20, 21]. mTORC2 is made up of mTOR, rapamycin-insensitive companion of mTOR (Rictor), protein observed with Rictor (Protor-1/2), mammalian stress-activated protein kinase interacting protein (mSIN1), mLST8, and DEPTOR [22]. Since some of the components

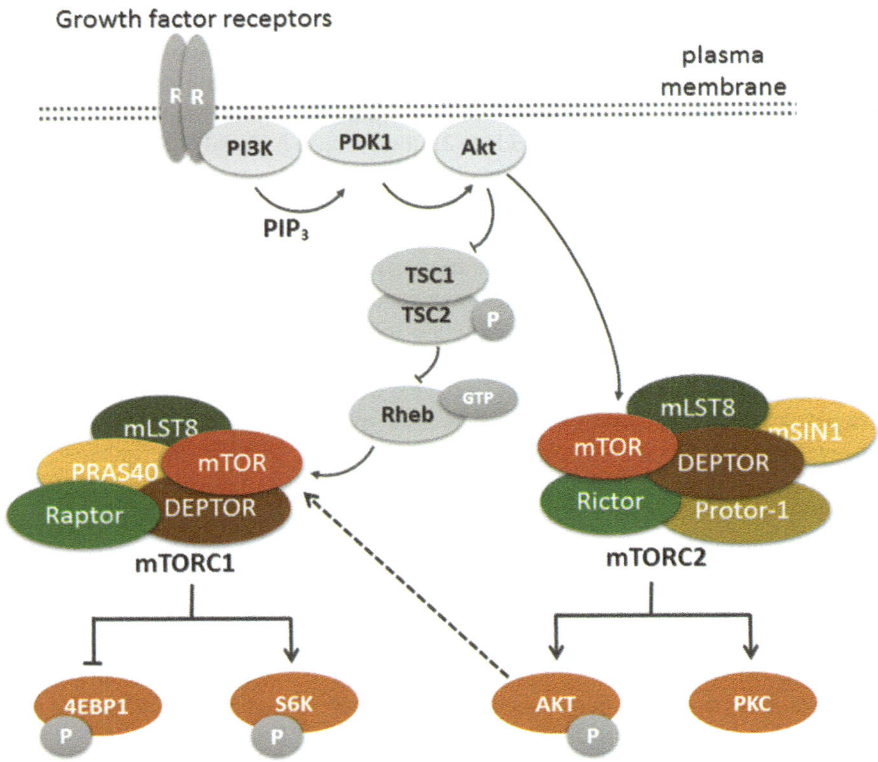


**Fig. 5.2** mTORC1 and mTORC2 complexes composition and signaling (see text for detailed explanation)

of these two complexes are the same, Raptor and Rictor are commonly used as markers to identify and distinguish mTORC1 and mTORC2, respectively. The functions of these two complexes are also different: mTORC1 is involved in regulation of cell growth, proliferation, protein and lipid biosynthesis, as well as regulation of autophagy, a lysosomal degradation pathway; while mTORC2, although less studied, is involved in cell survival, metabolism and cytoskeletal reorganization [15, 22]. Both complexes have been observed associated with various cellular compartments, such as lysosomes, mitochondria, nuclei, and the cytosol and it has been suggested that this localization is directly connected to mTOR function [23].

mTORC1 is activated by several factors. Activation by growth factors is a complicated and tightly controlled multistep process (Fig. 5.2) (reviewed in detail in [15, 23–25]). Briefly, growth factor/cytokine signaling through, for example, receptor tyrosine kinases, activates PI3K, phosphoinositide-dependent kinase-1 (PDK1), and AKT. AKT phosphorylates TSC2 on S939 and T1462 and thus inhibits the tuberous sclerosis TSC1/TSC2/TBC1D7 complex (TSC) [26, 27]. The TSC complex is a GTPase-activating protein (GAP) for the GTPase Ras homolog enriched in brain (Rheb) that functions as a negative regulator of mTORC1. Inhibition of the TSC complex allows Rheb-GTP to bind mTORC1 and promote its kinase activity [25]. Both Rheb and active mTORC1 are localized on the lysosomal surface.

mTORC1 is also activated by amino acids (reviewed in detail in [24, 28]). The exact mechanism is still being investigated, but so far it appears to involve several multiprotein complexes that regulate cellular responses to individual amino acids. In very simplified terms, in the presence of amino acids, active mTORC1 complex is located on the lysosome where it phosphorylates its substrates: eukaryotic translation initiation factor 4E-binding protein 1 (4EBP1) and ribosomal protein S6 kinase (S6K); both regulate downstream pathways necessary for protein and nucleotide synthesis. In the absence of amino acids, inactive mTORC1 has been reported to dissociate from the lysosome [29, 30]. Rheb and Rag GTPases, also located on the lysosomal surface, are necessary for mTORC1 activity. Several other multiprotein complexes, such as GATOR1 (GAP activity toward the Rag GTPases 1), GATOR2, KICSTOR (a scaffold for GATOR1) (reviewed in [15, 24]), have also been reported to regulate mTORC1 activity. Solute carrier family 38 member 9 (SLC38A9) and CASTOR1 have been described to serve as arginine sensors [31–33], while Sestrin1 and Sestrin2 have been identified as leucine sensors [34, 35].

In addition to all of these multiprotein complexes, active mTORC1 is tethered to the lysosome *via* Ragulator, a pentameric scaffolding complex that also anchors Rag GTPases to the lysosome. Furthermore, mTORC1 activation is directly linked to the V-ATPases: some of the subunits of the Ragulator complex directly interact with several V-ATPase subunits [30]. The V-ATPases are necessary for mTORC1 activation as inhibition of the V-ATPases using inhibitors or siRNA decreases mTORC1 activity; however, a precise role of the V-ATPases in mTORC1 signaling is not known [29, 30]. In the absence of amino acids, inactive mTORC1 allows initiation of autophagy, a lysosomal degradation process, to raise intracellular free amino acid levels by degrading proteins and organelles to survive this temporary "starvation" [36].

Less is known about mTORC2 signaling: mTORC2 is not activated by amino acids and is less sensitive to rapamycin treatment [22]. As described above, growth factor signaling activates AKT (S308), which, in turn phosphorylates mTOR on T2173 (in both mTORC1 and mTORC2 complexes) [37]. mTORC2 has been shown to phosphorylate AKT (S473) leading to maximal activation and stabilisation of AKT, thus connecting the mTORC1 and mTORC2 pathways (Fig. 5.2) [38].

## 5.3 Role of mTOR in Osteoclasts

Osteoclasts are unique cells: they are multinucleated (up to 20–30 nuclei per cell in pathological conditions); they contain numerous mitochondria; they have high levels of lysosomal membrane proteins and V-ATPases; during resorption, they secrete large amounts of proteolytic enzymes to degrade the demineralized bone matrix. All these processes require increased energy demands, as well as protein and lipid synthesis. The cellular master switch responsible for regulation of cell survival, cell proliferation, lipogenesis, protein synthesis, nucleotide synthesis, lysosome and mitochondrial biogenesis in all eukaryotic cells is, in fact, mTOR. Even though mTOR is involved in all of these cellular processes, surprisingly little is known about the precise role of mTOR in regulation of osteoclast differentiation and function. Below is the summary of what we do know so far.

During osteoclastogenesis, mTOR mRNA expression is increased at the pre-osteoclast stage, but returns to baseline levels in mature osteoclasts (our unpublished data and [39]). At the same time, gene expression levels of the mTORC1 and mTORC2 specific subunits Raptor and Rictor do not change over the course of osteoclastogenesis [39]. The activity of mTORC1, as measured by phosphorylation of S6K and S6, is also increased during the early/proliferation phase and then rapidly declines to almost undetectable levels in mature multinucleated osteoclasts [40, 41]. Both RANKL and M-CSF activate mTORC1 signaling, as determined by phosphorylation of mTORC1 substrates S6K, S6, and 4EBP1 [42]. mTOR has been shown to play a role in osteoclast survival: mTOR downregulates Bim (also known as BCL2-like protein 11), a proapoptotic BH3 domain only protein, and the decreased levels of Bim allow for osteoclast survival [43]. Treatment with mTORC1 inhibitor rapamycin or with mTOR siRNA inhibits osteoclast formation and induces apoptosis, confirming that mTORC1 is necessary for osteoclastogenesis and survival [42, 43]. Interestingly, it was observed that rapamycin had a more pronounced effect on osteoclast differentiation when cells were treated with the inhibitor at the earlier (days 1–2), rather than later (days 3–4) stages of osteoclastogenesis [39]. Furthermore, genetic deletion of mTOR or Raptor in vitro also significantly suppressed osteoclastogenesis in cells derived from bone marrow of mTOR *fl/–* or Raptor *fl/fl* mice [39]. Similar observations were also made in vivo: rapamycin treatment inhibited metastasis-induced osteoclastogenesis, as well as bone resorption [44]. These observations suggested that mTORC1 activity is more important at the precursor proliferation/early commitment stage rather than at the mature osteoclast stage.

Several in vivo studies have been published in the last 2 years (summarized in Table 5.1), which methodically deciphered and shed the light on the role of mTORC1 in osteoclast biology. To elucidate the role of mTORC1 in osteoclastogenesis, Wu et al. [45] created two osteoclast-specific conditional knockout mouse models by targeting the mTORC1 negative regulator TSC1 in osteoclast precursors (*LysM*-Cre;*Tsc1*$^{fl/fl}$ mice) or in mature osteoclasts (*Ctsk*-Cre;*Tsc1*$^{fl/fl}$ mice). Unexpectedly, hyper-activation of mTORC1 resulted in high bone density/osteopetrotic phenotype in both mouse lines; however, the underlying osteoclast defect was

**Table 5.1** Summary of the in vivo and in vitro phenotypes of the conditional knockout mouse models

| Gene K/O (*fl/fl*) | Stage of OC-genesis | Bone phenotype | OC # in vivo | OC # in vitro | Mechanism | Reference |
|---|---|---|---|---|---|---|
| TSC1 ↑↑mTORC1 | OC precursor (LysM-Cre) | Osteopetrosis; ↑bone volume, ↑trabecular number | Same OC# as in control ↓bone resorption (↓CTX, DPD) | ↓OC# and ↓bone resorption | ↓NF-κB signaling | Wu et al. [45] |
| TSC1 ↑↑mTORC1 | OC precursor (LysM-Cre) | Osteopetrosis ↑bone volume, ↑trabecular spacing; decreased thickness | ↑OC# ↓bone resorption (↓CTX) | ↑OC# ↓bone resorption | ↓NFATc1 ↓NF-κB p100 ↓NF-κB p105 ↓NF-κB p50 | Zhang et al. [40] |
| TSC1 ↑↑mTORC1 | Mature OC (Ctsk-Cre) | Osteopetrosis; ↑bone volume, ↑trabecular number, ↑trabecular thickness | ↑OC# ↓bone resorption (↓CTX, DPD) | ↑OC# ↓bone resorption | ↓actin ring structures formation | Wu et al. [45] |
| TSC1 ↑↑mTORC1 | Mature OC (Ctsk-Cre) | Osteopetrosis; ↑ bone volume, ↑trabecular number, ↑trabecular thickness | ↑OC# and ↓bone resorption (↓CTX) | ↑OC# ↓bone resorption | ↓actin ring, podosome belt, and ruffled border structures formation; ↓Rac1/Cdc42 activity/GTP binding | Xu et al. [41] |
| Raptor ↓↓mTORC1 | OC precursor (LysM-Cre) | Osteopenia ↓bone volume, ↓trabecular thickness, ↓trabecular number | ↑OC# ↑ bone resorption (↑CTX) | ↓OC# ↑OC size ↑bone resorption | ↑NFATc1 | Zhang et al. [40] |
| Raptor ↓↓mTORC1 | Mature OC (Ctsk-Cre) | Osteopetrosis; ↑bone volume, ↑trabecular thickness | ↓OC# | ↓OC# | ↓OC-specific gene expression | Dai et al. [47] |

different [45]. *LysM*-Cre;*Tsc1*$^{fl/fl}$ mice had normal weight and size, and the number of osteoclasts in vivo did not appear to be affected by the deletion; however, bone resorption parameters (lower serum C-terminal telopeptide (CTX) and urine deoxypyridinoline (DPD) levels) were decreased. In vitro, monocyte proliferation was increased, while the number of multinucleated TRAP-positive cells and bone resorption were significantly diminished. Gene expression of the differentiation markers (e.g., DC-STAMP, NFATc1, CtsK, TRAP) was decreased, mainly due to reduced NFATc1 and NF-κB activity, therefore, explaining the failure to form multinucleated cells [45]. The bone and osteoclast phenotype, as well as an inhibition of NFATc1 and NF-κB were also confirmed independently by another group [40]. At the same time, the *Ctsk*-Cre;*Tsc1*$^{fl/fl}$ mice also had normal weight and size; however, the number of osteoclasts in vivo was dramatically increased [45]. Bone resorption, as measured by CTX and DPD levels, was significantly decreased in these mice, suggesting impaired osteoclast function. In vitro, the number and size of the *Ctsk*-Cre;*Tsc1*$^{fl/fl}$ osteoclasts was increased; however, the bone resorbing function was decreased. The authors also reported that the number of ring-like actin structures in the bones of both mouse lines was diminished, and this defect appeared to be more pronounced in *Ctsk*-Cre;*Tsc1*$^{fl/fl}$ osteoclasts, suggesting that the impaired bone resorption was due to actin ring formation defect [41, 45]. Another group, also using osteoclasts derived from *Ctsk*-Cre;*Tsc1*$^{fl/fl}$ mice, showed that hyperactivation of mTORC1 in mature osteoclasts disturbed podosome belt/actin ring assembly, resulting in decreased bone resorption in vivo and in vitro [41]. Interestingly, treatment with low doses of rapamycin rescued podosome belt assembly and bone resorbing function both in vivo and in vitro, suggesting that low levels of mTORC1 activity are still required for proper osteoclast function. Xu et al. [41] further showed that this actin ring/podosome assembly defect was dependent on mTOR regulation of small GTPases Cdc42 and Rac1, the regulators of the actin cytoskeleton and the GTPases involved in osteoclast migration, formation of actin ring, podosome belt, and ruffled border [46]. The osteoclasts from *Ctsk*-Cre;*Tsc1*$^{fl/fl}$ mice had lower levels of Rac1/Cdc42 activity compared to controls and the authors proposed that mTORC1 is an upstream negative regulator of Rac1/Cdc42 [41].

Two groups generated osteoclast-specific conditional knockout mouse models where mTORC1 was inactivated by targeting Raptor, the unique scaffolding protein in mTORC1, in osteoclast precursors and in mature osteoclasts [40, 47]. Interestingly, the mouse models had different bone phenotypes: *LysM*-Cre;*Raptor*$^{fl/fl}$ mice had osteopenia [40], while *Ctsk*-Cre;*Raptor*$^{fl/fl}$ mice were osteopetrotic [47]. The *LysM*-Cre;*Raptor1*$^{fl/fl}$ mice had a reduced bone mass and a significantly higher number of osteoclasts in vivo, together with an elevated bone resorption rate (as measured by serum CTX levels). In vitro, osteoclastogenesis using the cells from the *LysM*-Cre;*Raptor1*$^{fl/fl}$ mice was increased, generating higher number of larger (5+ nuclei) cells; gene expression of osteoclast-specific genes was also upregulated, suggesting an acceleration of differentiation compared to the controls. In addition, the *LysM*-Cre;*Raptor1*$^{fl/fl}$ osteoclasts generated larger resorption lacunae, confirming the in vivo phenotype [40]. Since the protein levels of the transcription factors NFATc1 and NF-κB2 were increased, the authors proposed that the noncanonical

NF-κB2 and NFATc1 are negatively regulated by mTORC1 during osteoclastogenesis [40]. In comparison, when mTORC1 was inactivated in mature osteoclasts, the *Ctsk*-Cre;*Raptor*$^{fl/fl}$ mice had lower bone mass and decreased number of osteoclasts [47]. In vitro, the number of multinucleated *Ctsk*-Cre;*Raptor*$^{fl/fl}$ osteoclasts was also decreased, even though osteoclast progenitor proliferation was not affected. The authors demonstrated that the expression of a constitutively active form of S6K1 rescued the osteoclast phenotype in vitro, confirming that mTORC1 activity is necessary for proper osteoclast maturation and function [47].

In summary, these latest osteoclast-specific conditional knockout models clearly demonstrate that mTORC1 signaling plays a crucial role in osteoclast formation and function. What is apparent so far, is the fact that mTORC1 has different roles during different stages of osteoclastogenesis: high mTORC1 activity is necessary for early precursor proliferation phase, while low levels of mTORC1 activity are required for the later stages—osteoclast fusion, cytoskeletal reorganization/actin ring/ruffled border formation, and bone resorption (furthermore, mTORC1 was recently shown to play a role in determining osteoclast size, both in continuous osteoclast fusion and in fusion-independent cytoplasm growth [48]). At the moment, it is hard to reconcile all the observations into a simple consistent model, but it is clear that dysregulation of mTORC1 can potentially lead to osteopetrosis or to osteopenia when activated or repressed at the wrong time.

## 5.4 Regulation of mTOR in Osteoclasts

Based on the studies summarized in the previous section, it is clear that mTORC1 activity levels differ at different stages of osteoclastogenesis, with higher protein levels of mTOR/mTORC1 activity at the earlier stages, and lower protein levels/activity at the later stages. The mechanisms of mTORC1 regulation in osteoclasts are still not known; however, there are potential clues suggesting that in osteoclasts mTOR is regulated differently compared to other cell types and cell lines.

The majority of published studies indicate that the following factors are involved in mTORC1 regulation: (1) nutrient/amino acid status, with mTORC1 reported to localize to the surface of the lysosomes and to dissociate during starvation; (2) autophagy, where active mTORC1 suppresses autophagy, while inactive mTORC1 induces autophagy [49]; (3) the V-ATPase function, where inhibition of the V-ATPase downregulates mTORC1 activity [29, 30]; and (4) lysosomal positioning, where peripheral *vs.* perinuclear localization of the lysosomes dictates mTORC1 activity levels [50]. Lysosomes appear to play a central role in mTORC1 regulation and function: it is the place for mTORC1 activity, substrate recruitment and phosphorylation; furthermore, mTORC1 activity is regulated by intraluminal amino acids *via* an unknown inside-out mechanism ([30] and see detailed reviews in [25, 51, 52]). Meanwhile, mTORC1 is responsible for regulation of lysosomal biogenesis: active mTORC1 phosphorylates (and inactivates) transcription factor EB

(TFEB), the transcription factor considered to be a master regulator of lysosomal biogenesis [53]. This creates an interdependent relationship between mTORC1 and the lysosome: the lysosome regulates mTORC1 activity, while mTORC1 controls lysosome formation and function.

Our laboratory is interested in investigating the role of the lysosomal pH in osteoclast differentiation and signaling. One of the model systems we use is a mouse model with the R740S mutation in the V-ATPase *a*3 subunit, where an evolutionary conserved arginine involved in proton translocation across the membrane is replaced with serine [54]. The *a*3 R740S mutation does not affect protein expression, and the V-ATPase multisubunit complexes are assembled and targeted to the lysosome; however, the proton pumping is impaired [55, 56]. The *a*3 containing V-ATPases are preferentially expressed in osteoclasts and are localized to the lysosomes and to the ruffled border membrane [57, 58]. Due to this high expression level, the *a*3 R740S mutation significantly affects osteoclast bone resorption: homozygous (R740S/R740S) mice have severe osteopetrorickets, and heterozygous (+/R740S) animals have mild osteopetrosis [56, 59]. Lysosomal pH in osteoclasts with the R740S mutation is higher compared to the wild type (+/+) controls: pH ~6.3 in the R740S/R740S cells *vs*. pH ~5.7 in +/R740S cells *vs*. pH ~4.7 in +/+ controls [55, 60]. This gene-dosage effect makes the R740S cells a perfect model to study the role of lysosomal pH in osteoclast signaling.

During characterization of the +/R740S osteoclasts, we demonstrated that the these cells had decreased osteoclastogenesis due to accumulation of regulator of calcineurin 1 (RCAN1), an endogenous inhibitor of NFATc1, resulting in impaired NFATc1 nuclear translocation [55]. As RCAN1 protein levels in the cells are controlled by lysosomal degradation [61], we investigated autophagy, a lysosomal degradation process dependent on proper lysosomal function. We made three interesting observations: (1) osteoclastogenesis was severely impaired in R740S/R740S cells (Fig. 5.3); (2) autophagic flux was blocked in cells with the R740S mutation; and (3) mTOR protein levels and mTORC1 activity was increased in cells with the R740S mutation [62, 63]. The last observation appeared to contradict the current model of mTORC1 signaling, a model stating that active mTORC1 inhibits autoph-

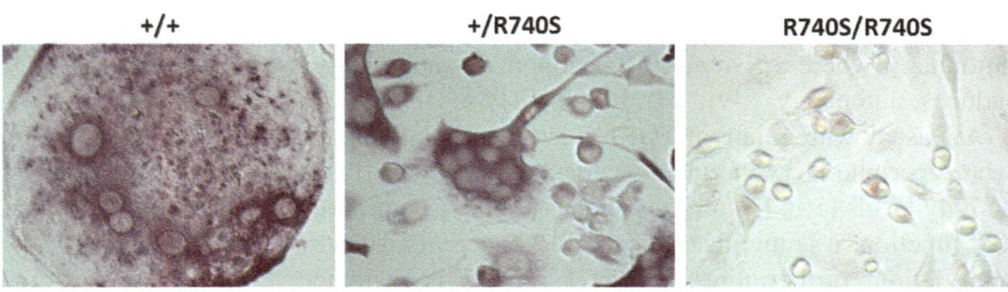


**Fig. 5.3** Osteoclastogenesis in cells with R740S mutation. Spleen-derived osteoclasts were differentiated in the presence of M-CSF and RANKL for 4 days. The cells were then fixed and stained for TRAP, an osteoclast marker. R740S/R740S cells had almost no cells with more than four nuclei (unpublished observations)

agy and, therefore, cannot coexist with active autophagic degradation. To verify our findings, we treated +/+ cells with the lysosomal inhibitors ammonium chloride ($NH_4Cl$) or chloroquine (CHQ) and confirmed that increased lysosomal pH resulted in higher mTOR protein levels and mTORC1 activity [62]. Based on our results, we hypothesized that in osteoclasts mTOR is regulated by lysosomal degradation. Treatment of +/+ and +/R740S osteoclasts with CHQ and proteasomal inhibitor MG132 increased mTOR protein levels in +/+ cells, but not in +/R740S osteoclasts, confirming our hypothesis. Cycloheximide blockade (inhibition of new protein synthesis) showed a decrease of mTOR protein levels in +/+ cells; however, the rate of the decrease in +/R740S cells was significantly slower, further supporting our lysosomal degradation hypothesis [62]. Our finding contradicting the current model of mTOR regulation is not unique and have been also observed in at least two other cell types: in primary mouse chondrocytes [64, 65] and in primary mouse hippocampal neurons [66]. Bartolomeo et al. reported that this increased mTORC1 activity was only observed in chondrocytes from the mouse model for mucopolysaccharidosis type VII (MPSVII), a lysosomal storage disease, but not in fibroblasts from MSPVII mice or HeLa cells lacking the same gene [64]. Furthermore, Hwang et al. showed that in ischemia-induced hippocampal neurons, mTOR was preferentially degraded *via* the autophagy/lysosomal pathway [66]. These results collectively suggest that mTOR/mTORC1 regulation by lysosomal degradation could be a special property of highly specialized cells, such as neurons, chondrocytes, and osteoclasts.

Lysosomal positioning is another factor reported to be involved in regulation of mTORC1 activity [50]. Using HeLa cells, Korolchuk et al. showed that during "starvation" (corresponding to inactive mTORC1), the lysosomes are located in the perinuclear region of the cells, while in the presence of the nutrients/amino acids (corresponding to active mTORC1), the lysosomes are dispersed in the cytosol and move toward cell periphery [50]. Furthermore, overexpression of factors that redistributed lysosomes to the periphery, e.g., kinesins KIF1Bβ and KIF2 and the small GTPase ARL8B, increased mTORC1 activity. Contrary to HeLa cells [50], in osteoclasts, lysosomes were primarily perinuclear during "fed" conditions, while "starvation" caused the lysosomes to move to the periphery [63]. Similar observations were reported for human adipose microvascular endothelial cells, primary human macrophages, and dendritic cells [67, 68], suggesting that different cell types have different pattern of lysosomal distribution. In addition to lysosomal distribution, we also observed that in osteoclasts mTOR does not disassociate from the lysosome during "starvation". Using an ultrapure lysosomal purification method, we demonstrated that absence of mTORC1/lysosome dissociation in the absence of nutrients was only observed in differentiated mature osteoclasts, but not in undifferentiated mouse monocyte cell line RAW264.7 [63].

In summary, we believe that mTOR regulation in osteoclasts (and possibly in other highly specialized cells, such as neurons) is different compared to other cell types and cell lines: (1) mTOR protein levels and mTORC1 activity appears to be regulated by lysosomal/autophagic degradation; (2) mTORC1 activity does not depend on lysosomal distribution; and (3) mTORC1 does not dissociate from the lysosome and remains associated with the lysosome even during "starvation." However, the exact mechanisms involved in mTORC1 regulation in osteoclasts are not known and still need to be elucidated.

## 5.5 Conclusion

Osteoclasts are unique bone-resorbing cells involved in maintaining bone homeostasis; however, increased osteoclast activity is responsible for pathological bone loss in numerous conditions, such as osteoporosis, osteoarthritis, rheumatoid arthritis, Paget's disease, and cancer-related osteolysis. As mTOR plays a key role in regulating osteoclast formation, activity, and function, mTORC1 signaling pathway could become a therapeutic target to treat diseases that involve overactive osteoclasts [38]. Since regulation of mTOR can be different in very specialized cells as we and others have shown, caution is necessary in extrapolating treatment paradigms from one cell/organ type to another.

## References

1. Crockett JC, et al. Bone remodelling at a glance. J Cell Sci. 2011;124(Pt 7):991–8.
2. Teti A. Mechanisms of osteoclast-dependent bone formation. Bonekey Rep. 2013;2:449.
3. Ono T, Nakashima T. Recent advances in osteoclast biology. Histochem Cell Biol. 2018;149:325.
4. Kong YY, et al. OPGL is a key regulator of osteoclastogenesis, lymphocyte development and lymph-node organogenesis. Nature. 1999;397(6717):315–23.
5. Hsu H, et al. Tumor necrosis factor receptor family member RANK mediates osteoclast differentiation and activation induced by osteoprotegerin ligand. Proc Natl Acad Sci U S A. 1999;96(7):3540–5.
6. Wiktor-Jedrzejczak W, et al. Total absence of colony-stimulating factor 1 in the macrophage-deficient osteopetrotic (op/op) mouse. Proc Natl Acad Sci U S A. 1990;87(12):4828–32.
7. Salo J, et al. Removal of osteoclast bone resorption products by transcytosis. Science. 1997;276(5310):270–3.
8. Tondravi MM, et al. Osteopetrosis in mice lacking haematopoietic transcription factor PU.1. Nature. 1997;386(6620):81–4.
9. Carey HA, et al. Enhancer variants reveal a conserved transcription factor network governed by PU.1 during osteoclast differentiation. Bone Res. 2018;6:8.
10. Yagi M, et al. DC-STAMP is essential for cell-cell fusion in osteoclasts and foreign body giant cells. J Exp Med. 2005;202(3):345–51.
11. Miyamoto H, et al. OC-STAMP and DC-STAMP cooperatively modulate cell-cell fusion to form osteoclasts and foreign body giant cells. J Bone Miner Res. 2012;27:1289.
12. McDonald M, et al. Intravital imaging of osteoclasts in vivo reveals cellular recycling as a novel cell fate mechanism. J Bone Miner Res. 2017;32(Suppl 1):Abstrac #1109.
13. Clausen BE, et al. Conditional gene targeting in macrophages and granulocytes using LysMcre mice. Transgenic Res. 1999;8(4):265–77.
14. Chiu WS, et al. Transgenic mice that express Cre recombinase in osteoclasts. Genesis. 2004;39(3):178–85.

15. Saxton RA, Sabatini DM. mTOR signaling in growth, metabolism, and disease. Cell. 2017;168(6):960–76.
16. Heitman J, Movva NR, Hall MN. Targets for cell cycle arrest by the immunosuppressant rapamycin in yeast. Science. 1991;253(5022):905–9.
17. Hara K, et al. Raptor, a binding partner of target of rapamycin (TOR), mediates TOR action. Cell. 2002;110(2):177–89.
18. Nojima H, et al. The mammalian target of rapamycin (mTOR) partner, raptor, binds the mTOR substrates p70 S6 kinase and 4E-BP1 through their TOR signaling (TOS) motif. J Biol Chem. 2003;278(18):15461–4.
19. Kim DH, et al. GbetaL, a positive regulator of the rapamycin-sensitive pathway required for the nutrient-sensitive interaction between raptor and mTOR. Mol Cell. 2003;11(4):895–904.
20. Peterson TR, et al. DEPTOR is an mTOR inhibitor frequently overexpressed in multiple myeloma cells and required for their survival. Cell. 2009;137(5):873–86.
21. Sancak Y, et al. PRAS40 is an insulin-regulated inhibitor of the mTORC1 protein kinase. Mol Cell. 2007;25(6):903–15.
22. Laplante M, Sabatini DM. Regulation of mTORC1 and its impact on gene expression at a glance. J Cell Sci. 2013;126(Pt 8):1713–9.
23. Betz C, Hall MN. Where is mTOR and what is it doing there? J Cell Biol. 2013;203(4):563–74.
24. Meng D, Frank AR, Jewell JL. mTOR signaling in stem and progenitor cells. Development. 2018;145(1):pii: dev152595.
25. Efeyan A, Zoncu R, Sabatini DM. Amino acids and mTORC1: from lysosomes to disease. Trends Mol Med. 2012;18(9):524–33.
26. Manning BD, et al. Identification of the tuberous sclerosis complex-2 tumor suppressor gene product tuberin as a target of the phosphoinositide 3-kinase/akt pathway. Mol Cell. 2002;10(1):151–62.
27. Han S, et al. Pam (Protein associated with Myc) functions as an E3 ubiquitin ligase and regulates TSC/mTOR signaling. Cell Signal. 2008;20(6):1084–91.
28. Wolfson RL, Sabatini DM. The dawn of the age of amino acid sensors for the mTORC1 pathway. Cell Metab. 2017;26(2):301–9.
29. Sancak Y, et al. Ragulator-Rag complex targets mTORC1 to the lysosomal surface and is necessary for its activation by amino acids. Cell. 2010;141(2):290–303.
30. Zoncu R, et al. mTORC1 senses lysosomal amino acids through an inside-out mechanism that requires the vacuolar H(+)-ATPase. Science. 2011;334(6056):678–83.
31. Wang S, et al. Metabolism. Lysosomal amino acid transporter SLC38A9 signals arginine sufficiency to mTORC1. Science. 2015;347(6218):188–94.
32. Rebsamen M, et al. SLC38A9 is a component of the lysosomal amino acid sensing machinery that controls mTORC1. Nature. 2015;519(7544):477–81.
33. Chantranupong L, et al. The CASTOR proteins are arginine sensors for the mTORC1 pathway. Cell. 2016;165(1):153–64.
34. Saxton RA, et al. Structural basis for leucine sensing by the Sestrin2-mTORC1 pathway. Science. 2016;351(6268):53–8.
35. Wolfson RL, et al. Sestrin2 is a leucine sensor for the mTORC1 pathway. Science. 2016;351(6268):43–8.
36. Shen HM, Mizushima N. At the end of the autophagic road: an emerging understanding of lysosomal functions in autophagy. Trends Biochem Sci. 2014;39(2):61–71.
37. Halova L, et al. Phosphorylation of the TOR ATP binding domain by AGC kinase constitutes a novel mode of TOR inhibition. J Cell Biol. 2013;203(4):595–604.
38. Perl A. Activation of mTOR (mechanistic target of rapamycin) in rheumatic diseases. Nat Rev Rheumatol. 2016;12(3):169–82.
39. Indo Y, et al. Metabolic regulation of osteoclast differentiation and function. J Bone Miner Res. 2013;28(11):2392–9.
40. Zhang Y, et al. mTORC1 inhibits NF-kappaB/NFATc1 signaling and prevents osteoclast precursor differentiation, in vitro and in mice. J Bone Miner Res. 2017;32(9):1829–40.

41. Xu S, et al. TSC1 regulates osteoclast podosome organization and bone resorption through mTORC1 and Rac1/Cdc42. Cell Death Differ. 2018;
42. Glantschnig H, et al. M-CSF, TNFalpha and RANK ligand promote osteoclast survival by signaling through mTOR/S6 kinase. Cell Death Differ. 2003;10(10):1165–77.
43. Sugatani T, Hruska KA. Akt1/Akt2 and mammalian target of rapamycin/Bim play critical roles in osteoclast differentiation and survival, respectively, whereas Akt is dispensable for cell survival in isolated osteoclast precursors. J Biol Chem. 2005;280(5):3583–9.
44. Hussein O, et al. Rapamycin inhibits osteolysis and improves survival in a model of experimental bone metastases. Cancer Lett. 2012;314(2):176–84.
45. Wu H, et al. Bone size and quality regulation: concerted actions of mTOR in mesenchymal stromal cells and osteoclasts. Stem Cell Rep. 2017;8(6):1600–16.
46. Touaitahuata H, Blangy A, Vives V. Modulation of osteoclast differentiation and bone resorption by Rho GTPases. Small GTPases. 2014;5:e28119.
47. Dai Q, et al. Inactivation of regulatory-associated protein of mTOR (raptor)/mammalian target of rapamycin complex 1 (mTORC1) signaling in osteoclasts increases bone mass by inhibiting osteoclast differentiation in mice. J Biol Chem. 2017;292(1):196–204.
48. Tiedemann K, et al. Regulation of osteoclast growth and fusion by mTOR/raptor and mTOR/rictor/Akt. Front Cell Dev Biol. 2017;5:54.
49. Noda T, Ohsumi Y. Tor, a phosphatidylinositol kinase homologue, controls autophagy in yeast. J Biol Chem. 1998;273(7):3963–6.
50. Korolchuk VI, et al. Lysosomal positioning coordinates cellular nutrient responses. Nat Cell Biol. 2011;13(4):453–60.
51. Puertollano R. mTOR and lysosome regulation. F1000Prime Rep. 2014;6:52.
52. Rabanal-Ruiz Y, Korolchuk VI. mTORC1 and Nutrient homeostasis: the central role of the lysosome. Int J Mol Sci. 2018;19(3):pii: E818.
53. Sardiello M, et al. A gene network regulating lysosomal biogenesis and function. Science. 2009;325(5939):473–7.
54. Kawasaki-Nishi S, Nishi T, Forgac M. Arg-735 of the 100-kDa subunit a of the yeast V-ATPase is essential for proton translocation. Proc Natl Acad Sci U S A. 2001;98(22):12397–402.
55. Voronov I, et al. The R740S mutation in the V-ATPase a3 subunit increases lysosomal pH, impairs NFATc1 translocation, and decreases in vitro osteoclastogenesis. J Bone Miner Res. 2013;28(1):108–18.
56. Ochotny N, et al. The V-ATPase a3 subunit mutation R740S is dominant negative and results in osteopetrosis in mice. J Bone Miner Res. 2011;26(7):1484–93.
57. Toyomura T, et al. Three subunit a isoforms of mouse vacuolar H(+)-ATPase. Preferential expression of the a3 isoform during osteoclast differentiation. J Biol Chem. 2000;275(12):8760–5.
58. Manolson MF, et al. The a3 isoform of the 100-kDa V-ATPase subunit is highly but differentially expressed in large (>or=10 nuclei) and small (<or= nuclei) osteoclasts. J Biol Chem. 2003;278(49):49271–8.
59. Ochotny N, et al. The R740S mutation in the V-ATPase a3 subunit results in osteoclast apoptosis and defective early-stage autophagy. J Cell Biochem. 2013;114(12):2823–33.
60. Johnson L, et al. V-ATPases containing a3 subunit play a direct role in enamel development in mice. J Cell Biochem. 2017;118(10):3328–40.
61. Liu H, et al. Degradation of regulator of calcineurin 1 (RCAN1) is mediated by both chaperone-mediated autophagy and ubiquitin proteasome pathways. FASEB J. 2009;23(10):3383–92.
62. Hu Y, et al. Lysosomal pH plays a key role in regulation of mTOR activity in osteoclasts. J Cell Biochem. 2016;117(2):413–25.
63. Wang A, et al. Activity-independent targeting of mTOR to lysosomes in primary osteoclasts. Sci Rep. 2017;7(1):3005.
64. Bartolomeo R, et al. mTORC1 hyperactivation arrests bone growth in lysosomal storage disorders by suppressing autophagy. J Clin Invest. 2017;127(10):3717–29.

65. Newton PT, et al. Pharmacological inhibition of lysosomes activates the MTORC1 signaling pathway in chondrocytes in an autophagy-independent manner. Autophagy. 2015;11(9):1594–607.
66. Hwang JY, et al. Global ischemia induces lysosomal-mediated degradation of mTOR and activation of autophagy in hippocampal neurons destined to die. Cell Death Differ. 2017;24:317.
67. Johnson DE, et al. The position of lysosomes within the cell determines their luminal pH. J Cell Biol. 2016;212(6):677–92.
68. Heuser J. Changes in lysosome shape and distribution correlated with changes in cytoplasmic pH. J Cell Biol. 1989;108(3):855–64.
69. Boyle WJ, Simonet WS, Lacey DL. Osteoclast differentiation and activation. Nature. 2003;423(6937):337–42.

# Chapter 6
# Apoptotic Cell Clearance in Gut Tissue: Role of Intestinal Regeneration

**Yasunao Numata, Daisuke Hirayama, Kohei Wagatsuma, Tomoya Iida, and Hiroshi Nakase**

**Abstract** Intestinal epithelial cells play a critical role in nutrient absorption as well as in protection against infection by pathogenic microorganisms. The cells drop out in a few days, and regeneration occurs subsequently; cells are eliminated by apoptosis. Clearance of dead cells frequently occurs in the intestinal tract, and apoptotic cells and phagocytes cooperate to facilitate cell clearance quickly and efficiently. The complex signaling network for cell clearance is well-understood. In recent years, the mechanism of programmed cell death accompanied by autophagy has been elucidated, and it has become clear that autophagy is involved in inflammation and intestinal tract diseases. In this review, we discuss intestinal regeneration and intestinal diseases through phagocytic clearance and autophagy of apoptotic cells.

## Abbreviations

| | |
|---|---|
| *ATG16L1* | Autophagy-related 16-like 1 |
| ATP | Adenosine triphosphate |
| C1q | Complement 1q |
| CD | Crohn's disease |
| CD14 | Cluster of differentiation 14 |
| CD31 | Cluster of differentiation 31 |
| CD36 | Cluster of differentiation 36 |
| CD47 | Cluster of differentiation 47 |
| CD91 | Cluster of differentiation 91 |
| CX3CL1 | Chemokine, CX3C motif, ligand 1 |
| EGF | Epidermal growth factor |
| GTPase | Guanosinetriphosphatase |
| GWAS | Genome-wide association studies |
| ICAM3 | Including intercellular adhesion molecule 3 |

Y. Numata · D. Hirayama · K. Wagatsuma · T. Iida · H. Nakase (✉)
Department of Gastroenterology and Hepatology, Sapporo Medical University School of Medicine, Sapporo, Hokkaido, Japan



K. Turksen (ed.), *Autophagy in Health and Disease*, Stem Cell Biology and Regenerative Medicine, https://doi.org/10.1007/978-3-319-98146-8_6

| | |
|---|---|
| IEL | Intraepithelial lymphocyte |
| LDL | Low density lipoprotein |
| LPC | Lysophosphatidylcholine |
| LRP1 | Low density lipoprotein receptor-related protein 1 |
| MFGE8 | Milk fat globule-EGF-factor 8 |
| *NOD2* | Nucleotide-binding oligomerization domain 2 |
| PtdSer | Phosphatidylserine |
| RGD | Arginine-Glycine-Aspartic Acid |
| ROS | Reactive oxygen species |
| S1P | Sphingosine-1-phosphate |
| TGF | Transforming growth factor |
| THP-1 | Tamm–Horsfall glycoprotein-1 |
| TIM1 | T-cell immunoglobulin and mucin containing protein-1 |
| TIM4 | T-cell immunoglobulin and mucin containing protein-4 |
| UTP | Uridine triphosphate |

## 6.1 Introduction

Apoptosis is known as programmed cell death, and it occurs in all tissues as part of normal development, homeostasis, and pathogenic processes [1]. Cells that become unnecessary for living bodies undergo apoptosis and are selectively cleared by phagocytes. Unlike necrosis, clearance of apoptotic cells by phagocytes (referred to as "efferocytosis") generally maintains immune tolerance and does not cause inflammation [2]. Therefore, apoptosis is an important mechanism for maintaining the homeostasis of living bodies and for tissue repair. In recent years, abnormalities of interstitial cell clearance have been considered as either a cause or a result of many diseases [3–6].

In contrast, autophagy is an intracellular degradative mechanism that decomposes intracellular components using the lysosome [7]. When intracellular organelles are destroyed, the proteins of the organelles are decomposed and metabolized in the cell; homeostasis and normal function are maintained. However, when cell organelle proteins cannot be completely decomposed in the cell, proteins accumulate and aggregate, causing the cell to commit suicide. Autophagy is recognized as a mechanism of programmed cell death different from apoptosis.

This review introduces the processing mechanisms of apoptotic cells, discusses its role in the regeneration of the intestinal tract, and describes an outline of autophagy and its relation to intestinal diseases.

## 6.2 Migration of Phagocytes to Apoptotic Cells

For the proper clearance of apoptotic cells by phagocytes, phagocytes need to be properly translocated to the dying cell. For this to occur, it is necessary for phagocytes to migrate toward apoptotic cells and distinguish them from their living neighbors [8].

When apoptosis occurs, the caspase cascade is activated, and soluble chemoattractants, called the "find-me signal," are released [9, 10]. These signals can diffuse through submicron vesicle membranes. Phagocytes are attracted by the signals and are properly translocated to the apoptotic cells, depending on the signal. It has been shown that the find-me signal enhances phagocyte localization to dead cells [11].

There are at least four distinct find-me signal-receptor pathways identified primarily through in vitro studies; these are mediated by various molecules, including the extracellular nucleotides ATP and UTP, CX3CL1, sphingosine-1-phosphate (S1P), and LysoPC (LPC) (see Fig. 6.1). The first three have been reported to be associated with efferocytosis in vivo [12–16].

ATP and UTP have been identified as find-me signal mediators in vitro and in vivo. For example, it has been reported that they induce strong chemotactic activity in THP-1(a human leukemia monocytic cell line) [12]; these factors generate a concentration gradient. The P2Y2 receptor detects the find-me signal, inducing subsequent phagocytosis and drawing the phagocyte toward the apoptotic cell [12]. Another example of a find-me signal mediator released from apoptotic cells is CX3CL1 (also known as "fractalkine"). Intercellular adhesion molecule 3 (ICAM3) and a proteolytically processed form of CX3CL1 also prompt macrophage migration toward apoptotic cells [16, 17]. S1P is also a potent chemoattractant for monocytes and macrophages [13, 15, 18, 19]. Nucleotides (ATP and UTP), CX3CL1, and S1P are released early by apoptotic lymphocytes, making apoptotic cell clearance rapid and efficient in tissues.

When apoptosis occurs, find-me signals are released from apoptotic cells. ATP/UTP, CX3CL1, S1P, and LPC have been identified as find-me signal mediators. These molecules bind to their specific receptors present on the surface of phagocytes. Phagocytes are activated by their stimulation and migrate toward the dead cells.

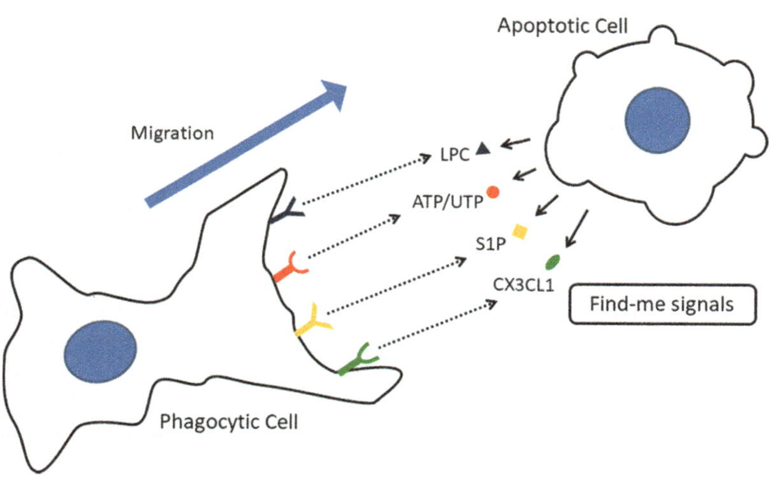


**Fig. 6.1** Find-me signals and migration of phagocytes

## 6.3 Recognition of Apoptotic Cells

Apoptotic cells release the find-me signal to be recognized by the recruited phagocyte. In addition, they express a specific molecule, the "eat-me signal," on the cell surface to promote their recognition by the phagocyte. Among eat-me signals, phosphatidylserine (PtdSer) is the most well-known [8].

Effects of enzymes such as aminophospholipid translocase have preserved the asymmetry of the phospholipids in the cell membrane. When these enzyme activities are inhibited in the early stages of apoptosis, PtdSer is expressed on the cell surface [2]. This increased exposure of PtdSer on the surface of the cell, thought as a unique feature of apoptotic cells, is recognized by the phagocytic PtdSer receptor and functions to trigger of phagocytosis [20].

However, it seems that the expression of certain eat-me signals alone is insufficient to cause the phagocytosis of apoptotic cells [21, 22].

In contrast to the eat-me signals, there are also "don't-eat-me" signals. Some surface proteins, such as adhesion-related proteins CD47 and CD31, act as signals to suppress phagocytosis [11].

It is reported that apoptotic cell clearance by phagocytosis requires both the sufficient expression of an eat-me signal on the surface of an apoptotic cell, and the absence of a don't-eat-me signal [1].

## 6.4 Phagocytosis of Apoptotic Cells

On the surface of phagocytes, receptors are expressed excessively to recognize eat-me signals on the surface of apoptotic cells. These play a role to connect phagocytes and apoptotic cells (tethering) [2].

The integrin $\alpha_v\beta_{3/5}$ is an apoptotic cell receptor expressed on phagocytes. It binds MFGE8, a PtdSer binding protein, to phagocytes via its RGD motif. Then, this complex binds PtdSer on apoptotic cells [5, 23–27]. It is reported that receptors such as BAI1, TIM-4, and TIM-1 recognize PtdSer on apoptotic cells directly [28–31].

Aside from lectins that bind modified sugars on apoptotic cells [32], CD36 (in conjunction with integrins $\alpha_v\beta_3$ and $\alpha_v\beta_5$) that binds thrombospondin [27], LRP1/CD91 (in conjunction with calreticulin) that binds complement component C1q [33], CD14 that binds ICAM3 [34], and the scavenger receptors that bind oxidized LDL [35] have all been identified as receptors expressed on the surface of phagocytes.

However, phagocytes do not take up apoptotic cells at this stage. When PtdSer receptors are involved (thickling), cellular uptake is initiated [2]. When phagocytes capture apoptotic targets, actin filaments within the phagocytes assemble and separate, the cytoskeleton is rearranged, and apoptotic cells are taken up in the membrane. The engulfment of apoptotic cells is regulated by Rho family GTPases (Rac1, RhoA, Rab5, etc.) [36]. When RhoA is deleted or downregulated in phagocytes, phagocytosis is enhanced; on the other hand, when RhoA is upregulated, phagocytosis is suppressed. RhoA activation is regulated by the acidification of phagocytes [37]. When phagocytes begin to engulf apoptotic cells, Rac1 and integrins activate the formation of the phagocytic cup made of actin filaments. As soon as apoptotic cells are taken up by the cup of the phagocyte, Rac1 is inactivated, and actin is depolymerized. Then, the components of the dead cell are transferred to the lysosome, mediated by Rab5 [38, 39].

In another pathway, actin polymerization in phagocytes is regulated. This pathway involves CrkII, Dock180, and ELMO [40–44], which activate Rac-GTP. The pathway is thought to be involved in apoptotic cell engulfment, cell migration, and neurite growth. However, the upstream receptors involved in the activation of this signaling pathway are still unknown (see Fig. 6.2).

Apoptotic cells express eat-me signals on the surface to promote their recognition by the phagocyte. Phosphatidylserine (PtdSer) is the most well-known among eat-me signals. In the early stages of apoptosis, apoptotic cells expose inner-membrane PtdSer on the surface. When phagocytes start to engulf apoptotic cells, phagocytosis is regulated by the Rho-GTPase-dependent pathway; actin polymerization and phagocyte migration are regulated by the Rac-GTPase-dependent pathway.

## 6.5 Cell Clearance and Anti-inflammatory Response

Clearance of apoptotic cells is initiated from the earliest stages of apoptosis to avoid the release of cell contents upon the collapse of the cell membrane. However, when cell membrane integrity is lost, necrosis ensues, cell contents are released, and the immune response is activated [45]. If proper recognition, clearance, and degradation of the apoptotic cellular material is not carried out, autoantigens are exposed, causing an autoimmune response and uncontrolled inflammation to occur [39, 46, 47]. To maintain an environment protecting the host from autoimmunity, rapid clearance of apoptotic cells is essential. The clearance of early-stage apoptotic cells occurs and anti-inflammatory mediators such as IL-10 and TGF-β are released to decrease the severity of inflammation [48]. In contrast, the retardation or impairment of clearance of apoptotic cells may exacerbate inflammation and result in inflammatory diseases. Recent studies suggest that apoptotic cells are major factors in the presentation of autoantigens, causing systemic autoimmunity in susceptible hosts [49].

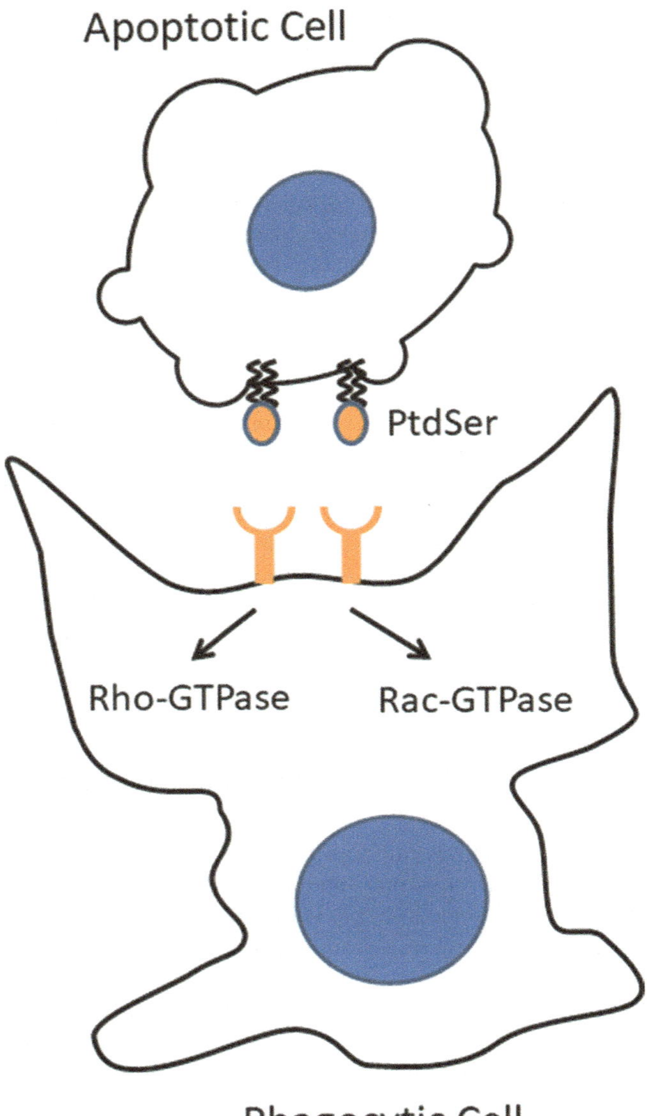


**Fig. 6.2** Phosphatidylserine as an Eat-me signal and Phagocytosis

## 6.6 Exclusion and Regeneration of Intestinal Epithelial Cells

Apoptosis of intestinal epithelial cells under physiological conditions occurs at the tips of villi. Macrophages and intraepithelial lymphocytes induce apoptosis; TNF-α, released from macrophages, is known to induce apoptosis in epithelial cells, in particular [50, 51]. T lymphocytes in the intestinal epithelium are called intraepithelial lymphocyte (IEL). The activation of IEL in vivo has been shown to induce epithelial apoptosis, eventually leading to mucosal damage [52]. IEL is thought to induce apoptosis through cytotoxic substances or through physical attack as a result of cellular processes [51].

Intestinal epithelia have many opportunities to contact apoptosis-inducing factors (toxins, bacteria, virus, etc.), and the strong action of these external factors can lead to the breakdown of the epithelial barrier and damage to the mucous membrane. Epithelial cells are therefore tightly closed with cell adhesion mechanisms. The barrier function provided by cell adhesion mechanisms is necessary; when cells escape from the epithelial line, they fall off the epithelium, and apoptosis is thought to ensue upon detachment [53]. In mice, a-zipper like cell adhesion mechanism is responsible for maintaining barrier function. In guinea pigs, epithelial cells are phagocytosed by lamina propria macrophages. Microvilli remain attached to the cytoplasm on the luminal side; this area is associated with normal epithelial cells on both sides, with the barrier function maintained. Neighboring epithelial cells subsequently come into contact and create assemblies that facilitate cell adhesion mechanisms. The epithelium that was sloughed off falls into the lumen [53, 54].

Thus, intestinal tract apoptosis and regeneration are necessary processes in maintaining barrier function.

## 6.7 Autophagy

Regarding physiological programmed cell death, there is also a mechanism for degrading intracellular proteins called "autophagy." The big picture behind the molecular mechanisms of apoptosis and its physiological significance is currently being revealed. In contrast, there is still so much to be discovered regarding the mechanisms behind autophagy-based cell death. Autophagy is a bulk protein degradation system for digesting cytoplasmic contents; it is a normal cell function essential for survival.

Autophagy refers to any cellular degradative pathway that involves the delivery of cytoplasmic cargo to the lysosome [55]. In this process, the endoplasmic reticulum or other membranous cellular structures generate a double-membrane structure called a "phagophore," this elongates to envelop the cytoplasmic contents or organelles to be degraded, forming an "autophagosome," a unique double-membrane organelle. The outer membrane of the autophagosome fuses with a lysosome to form an "autolysosome," and the inner membrane degrades and absorbs its contents [56–58]. The whole process is involves more than 30 kinds of autophagy-related genes and proteins (see Fig. 6.3) [59, 60].

Autophagy plays roles not only in facilitating the degradation of damaged cellular components, but also in stress response. Starved cells degrade their own proteins into amino acids and either transform them into energy sources, or use them as raw materials to produce new proteins. Autophagy is also an important mechanism for cells dealing with pathogenic infections [61].

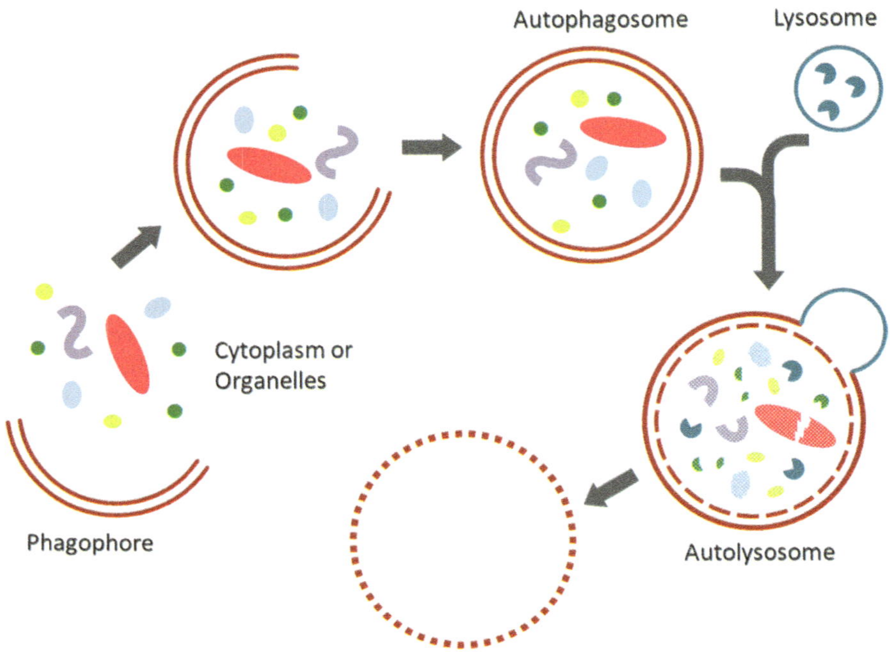


**Fig. 6.3** Schematic diagram of Autophagy

It has been reported that autophagy is associated with apoptosis, inflammation, immune reaction, ROS stress, and mitochondrial dysfunction. Autophagy is an important factor in the pathogenesis and regulation of various types of inflammation and immune related diseases [62–65]. Currently, autophagy is recognized to be involved in many diseases such as cancer, neurodegeneration, autoimmune diseases like CD, rheumatoid arthritis, and infection [61].

The process of autophagy begins with the formation of a double-membrane structure which is generated from the endoplasmic reticulum or from other membranous cellular structures. The double-membrane structure is known as the phagophore. It elongates to envelop the cytoplasmic contents organelles to be degraded and forms a double-membraned organelle, the autophagosome. Then the outer membrane of autophagosome fuses with the lysosome to generate the autolysosome, inside of which the sequestered cargo is degraded.

## 6.8 Intestinal Self-Defense and Immune Response

The intestinal tract houses a great diversity of microorganisms; thus, the host has to maintain a stable coexistence with them. However, the gut must still appropriately uphold immunity against pathogenic microorganisms that may arise [66, 67]. In general, the intestinal mucosal barrier is composed of two layers, the intestinal mucus layer and the epithelium. The intestinal mucus layer is covered with a layer of sticky gel [68, 69].

The epithelium is mostly composed of several kinds of intestinal epithelial cells, including absorptive enterocytes, goblet cells, enteroendocrine cells, and Paneth cells [70].

In the large intestine, the mucus layer covering the intestinal epithelial cells is a physical barrier against bacterial invasion into the epithelium. Furthermore, recent research has revealed that the protein Lypd8 mediates the separation of intestinal bacteria and epithelia cells by binding to flagella and suppressing the motility of flagellated bacteria [71].

Paneth cells are specialized epithelial cells that are involved in the innate immunity in the small intestine. When they come into contact with bacteria or other immunogens, these cells release secretory granules containing antimicrobial peptides and a variety of proteins to prevent pathogenic infection [62]. In recent years, researchers have discovered that autophagy and autophagy-related genes have a specific role in the physiology of Paneth cells [72]. Paneth cells are mainly localized in the small intestine; therefore Paneth cells are implicated in the development of inflammatory bowel diseases, particularly CD [73].

Currently, many genome-wide association studies (GWAS) have been performed, and over 100 genes have been identified to influence susceptibility to CD [74]. *NOD2* was the first disease-susceptibility gene discovered for CD [75]. The *NOD2* protein is responsible for immunity against intracellularly bacterial infection by associating with *ATG16L1*, an important molecule in the autophagy pathway, and mediating autophagy.

Mutant *NOD2* resulted in the impaired encapsulation of invading bacteria by autophagosomes. In this regard, CD patients with mutant *NOD2* are considered to exhibit disorders of autophagy. *ATG16L1* was also identified as a disease-susceptibility gene of CD; it encodes autophagy-related proteins. In addition to the decomposition of intracellular organelles, *ATG16L1* is involved in the degradation and elimination of bacteria, viruses, and other infectious substances. It is reported that the incidence of CD is twice as high in the *ATG16L1* mutants; this is because the level of occurrence of autophagy in the Paneth cells of *ATG16L1* mutant has been altered, inhibiting the formation of normal secretory granules [73, 76].

The relationship between autophagy and CD is currently being actively researched around the world. The results of future research are expected to guide the development of new policies for the treatment and prevention of CD.

## 6.9 Summary

Apoptotic cell clearance is an important process that helps to suppress excessive immunological response and to maintain homeostasis. In this review, we have outlined and discussed the steps of the cell clearance process, the function of find-me signals for locating phagocytes on apoptotic cells properly, the recognition of apoptotic cells via PtdSer, and the regulation of phagocytes. We have also mentioned apoptosis and intestinal regeneration in intestinal epithelial cells, as well as the relationship between autophagy and intestinal tract diseases. A better understanding of autophagy can further shed light on the pathophysiology of several diseases and the development of new therapies.

## References

1. Poon IK, Lucas CD, Rossi AG, Ravichandran KS. Apoptotic cell clearance: basic biology and therapeutic potential. Nat Rev Immunol. 2014;14(3):166–80.
2. Ogawa N, Shimoyama K, Kawanami T. Apotosisu Saibou no Kuriaransu to Jikomenneki (Clearance of apoptotic cells and autoimmunity). Nihon Rinsho. 2005;63(5):229–32.
3. Elliott MR, Ravichandran KS. Clearance of apoptotic cells: implications in health and disease. J Cell Biol. 2010;189(7):1059–70.
4. Green DR, Oguin TH, Martinez J. The clearance of dying cells: table for two. Cell Death Differ. 2016;23(6):915–26.
5. Nagata S, Hanayama R, Kawane K. Autoimmunity and the clearance of dead cells. Cell. 2010;140(5):619–30.
6. Saas P, Kaminski S, Perruche S. Prospects of apoptotic cell-based therapies for transplantation and inflammatory diseases. Immunotherapy. 2013;5(10):1055–73.
7. Ke P, Shao BZ, Xu ZQ, Chen XW, Liu C. Intestinal autophagy and its pharmacological control in inflammatory bowel disease. Front Immunol. 2016;7:695.
8. Chekeni FB, Ravichandran KS. The role of nucleotides in apoptotic cell clearance: implications for disease pathogenesis. J Mol Med (Berl). 2011;89(1):13–22.
9. Lauber K, Blumenthal SG, Waibel M, Wesselborg S. Clearance of apoptotic cells: getting rid of the corpses. Mol Cell. 2014;14(3):277–87.
10. Ravichandran KS, Lorenz U. Engulfment of apoptotic cells: signals for a good meal. Nat Rev Immunol. 2007;7(12):964–74.
11. Elliott MR, Ravichandran KS. The dynamics of apoptotic cell clearance. Dev Cell. 2016;38(2):147–60.
12. Elliott MR, Chekeni FB, Trampont PC, Lazarowski ER, Kadl A, Walk SF, Park D, Woodson RI, Ostankovich M, Sharma P, Lysiak JJ, Harden TK, Leitinger N, Ravichandran KS. Nucleotides released by apoptotic cells act as a find-me signal to promote phagocytic clearance. Nature. 2009;461(7261):282–6.
13. Gude DR, Alvarez SE, Paugh SW, Mitra P, Yu J, Griffiths R, Barbour SE, Milstien S, Springel S. Apoptosis induces expression of sphingosine kinase 1 to release sphingosine-1-phosphate as a "come-and-get-me" signal. FASEB J. 2008;22(8):2629–38.
14. Lauber K, Bohn E, Kröber SM, Xiao YJ, Blumenthal SG, Lindemann RK, Marini P, Wiedig C, Zobywalski A, Baksh S, Xu Y, Autenrieth IB, Schulze-Osthoff K, Belka C, Stuhler G, Wesselborg S. Apoptotic cells induce migration of phagocytes via caspase-3-mediated release of a lipid attraction signal. Cell. 2003;113(6):717–30.
15. Luo B, Gan W, Liu Z, Shen Z, Wang J, Shi R, Liu Y, Liu Y, Jiang M, Zhang Z, Wu Y. Erythropoeitin signaling in macrophages promotes dying cell clearance and immune tolerance. Immunity. 2016;44(2):287–302.
16. Truman LA, Ford CA, Pasikowska M, Pound JD, Wilkinson SJ, Dumitriu IE, Melville L, Melrose LA, Ogden CA, Nibbs R, Graham G, Combadiere C, Gregory CD. CX3CL1/fractalkine is released from apoptotic lymphocytes to stimulate macrophage chemotaxis. Blood. 2008;112(13):5026–36.

17. Torr EE, Gardner DH, Thomas L, Goodall DM, Bielemeier A, Willetts R, Griffiths HR, Marshall LJ, Devitt A. Apoptotic cell-derived ICAM-3 promotes both macrophage chemoattraction to and tethering of apoptotic cells. Cell Death Differ. 2012;19(4):671–9.
18. Weigert A, Cremer S, Schmidt MV, von Knethen A, Angioni C, Geisslinger G, Brüne B. Cleavage of sphingosine kinase 2 by caspase-1 provokes its release from apoptotic cells. Blood. 2010;115(17):3531–40.
19. Weigert A, Johann AM, von Knethen A, Schmidt H, Geisslinger G, Brüne B. Apoptotic cells promote macrophage survival by releasing the antiapoptotic mediator sphingosine-1-phosphate. Blood. 2006;108(5):1635–42.
20. Fadok VA, Voelker DR, Campbell PA, Cohen JJ, Bratton DL, Henson PM. Exposure of phosphatidylserine on the surface of apoptotic lymphocytes triggers specific recognition and removal by macrophages. J Immunol. 1992;148(7):2207–16.
21. Gardai SJ, McPhillips KA, Frasch SC, Janssen WJ, Starefeldt A, Murphy-Ullrich JE, Bratton DL, Oldenborg PA, Michalak M, Henson PM. Cell-surface calreticulin initiates clearance of viable or apoptotic cells through trans-activation of LRP on the phagocyte. Cell. 2005;123(2):321–34.
22. Segawa K, Suzuki J, Nagata S. Constitutive exposure of phosphatidylserine on viable cells. Proc Natl Acad Sci U S A. 2011;108(48):19246–51.
23. Hanayama R, Tanaka M, Miwa K, Shinohara A, Iwamatsu A, Nagata S. Identification of a factor that links apoptotic cells to phagocytes. Nature. 2002;417(6885):182–7.
24. Hanayama R, Tanaka M, Miyasaka K, Aozasa K, Koike M, Uchiyama Y, Nagata S. Autoimmune disease and impaired uptake of apoptotic cells in MFG-E8-deficient mice. Science. 2004;304(5674):1147–50.
25. Hynes RO. Integrins: bidirectional, allosteric signaling machines. Cell. 2002;110(6):673–87.
26. Overstreet MG, Gaylo A, Angermann BR, Hughson A, Hyun YM, Lambert K, Acharya M, Billroth-Maclurg AC, Rosenberg AF, Topham DJ, Yagita H, Kim M, Lacy-Hulbert A, Meier-Schellersheim M, Fowell DJ. Inflammation-induced interstitial migration of effector $CD4^+$ T cells is dependent on integrin αV. Nat Immunol. 2013;14(9):949–58.
27. Savill J, Dransfield I, Hogg N, Haslett C. Vitronectin receptor-mediated phagocytosis of cells undergoing apoptosis. Nature. 1990;343(6254):170–3.
28. Park D, Tosello Trampont A-C, Elliott MR, Lu M, Haney LB, Ma Z, Klibanov AL, Mandell JW, Ravichandran KS. BAI1 is an engulfment receptor for apoptotic cells upstream of the ELMO/Dock180/Rac module. Nature. 2007;450(7168):430–4.
29. Kobayashi N, Karisola P, Peña Cruz V, Dorfman DM, Jinushi M, Umetsu SE, Butte MJ, Nagumo H, Chernova I, Zhu B, Sharpe AH, Ito S, Dranoff G, Kaplan GG, Casasnovas JM, Umetsu DT, Dekruyff RH, Freeman GJ. TIM-1 and TIM-4 glycoproteins bind phosphatidylserine and mediate uptake of apoptotic cells. Immunity. 2007;27(6):927–40.
30. Miyanishi M, Tada K, Koike M, Uchiyama Y, Kitamura T, Nagata S. Identification of Tim4 as a phosphatidylserine receptor. Nature. 2007;450(7168):435–9.
31. Santiago C, Ballesteros A, Martínez Muñoz L, Mellado M, Kaplan GG, Freeman GJ, Casasnovas JM. Structures of T cell immunoglobulin mucin protein 4 show a metal-Ion-dependent ligand binding site where phosphatidylserine binds. Immunity. 2007;27(6):941–51.
32. Ezekowitz RA, Sastry K, Bailly P, Warner A. Molecular characterization of the human macrophage mannose receptor: demonstration of multiple carbohydrate recognition-like domains and phagocytosis of yeasts in Cos-1 cells. J Exp Med. 1990;172(6):1785–94.
33. Ogden CA, Decathelineau A, Hoffmann PR, Bratton D, Ghebrehiwet B, Fadok VA, Henson PM. C1q and mannose binding lectin engagement of cell surface calreticulin and CD91 initiates macropinocytosis and uptake of apoptotic cells. J Exp Med. 2001;194(6):781–95.
34. Gregory CD, Devitt A, Moffatt O. Roles of ICAM-3 and CD14 in the recognition and phagocytosis of apoptotic cells by macrophages. Biochem Soc Trans. 1998;26(4):644–9.
35. Gordon S. Macrophage-restricted molecules: role in differentiation and activation. Immunol Lett. 1999;65(1-2):5–8.

36. Nakaya M, Tanaka M, Okabe Y, Hanayama R, Nagata S. Opposite effects of rho family GTPases on engulfment of apoptotic cells by macrophages. J Biol Chem. 2006;281(13):8836–42.
37. Erwig LP, Henson PM. Clearance of apoptotic cells by phagocytes. Cell Death Differ. 2008;15(2):243–50.
38. Kitano M, Nakaya M, Nakamura T, Nagata S, Matsuda M. Imaging of Rab5 activity identifies essential regulators for phagosome maturation. Nature. 2008;453(7192):241–5.
39. Hochreiter Hufford A, Ravichandran KS. Clearing the dead: apoptotic cell sensing, recognition, engulfment, and digestion. Cold Spring Harb Perspect Biol. 2013;5(1):a008748.
40. Gumienny TL, Brugnera E, Tosello Trampont AC, Kinchen JM, Haney LB, Nishiwaki K, Walk SF, Nemergut ME, Macara IG, Francis R, Schedl T, Qin Y, Van Aelst L, Hengartner MO, Ravichandran KS. CED-12/ELMO, a novel member of the CrkII/Dock180/Rac pathway, is required for phagocytosis and cell migration. Cell. 2001;107(1):27–41.
41. Reddien PW, Horvitz HR. CED-2/CrkII and CED-10/Rac control phagocytosis and cell migration in Caenorhabditis elegans. Nat Cell Biol. 2000;2(3):131–6.
42. Wu YC, Horvitz HR. C. elegans phagocytosis and cell-migration protein CED-5 is similar to human DOCK180. Nature. 1998;392(6675):501–4.
43. Zhou Z, Caron E, Hartwieg E, Hall A, Horvitz HR. The C. elegans PH domain protein CED-12 regulates cytoskeletal reorganization via a Rho/Rac GTPase signaling pathway. Dev Cell. 2001;1(4):477–89.
44. Wu YC, Tsai MC, Cheng LC, Chou CJ, Weng NY. C. elegans CED-12 acts in the conserved crkII/DOCK180/Rac pathway to control cell migration and cell corpse engulfment. Dev Cell. 2001;1(4):491–502.
45. Arandjelovic S, Ravichandran KS. Phagocytosis of apoptotic cells in homeostasis. Nat Immunol. 2015;16(9):907–17.
46. Franz S, Gaipl US, Munoz LE, Sheriff A, Beer A, Kalden JR, Herrmann M. Apoptosis and autoimmunity: when apoptotic cells break their silence. Curr Rheumatol Rep. 2006;8(4):245–7.
47. Janko C, Franz S, Munoz LE, Siebig S, Winkler S, Schett G, Lauber K, Sheriff A, van der Vlag J, Herrmann M. CRP/anti-CRP antibodies assembly on the surfaces of cell remnants switches their phagocytic clearance toward inflammation. Front Immunol. 2011;2:70.
48. Pasparakis M, Vandenabeele P. Necroptosis and its role in inflammation. Nature. 2015;517(7534):311–20.
49. Casciola Rosen LA, Anhalt G, Rosen A. Autoantigens targeted in systemic lupus erythematosus are clustered in two populations of surface structures on apoptotic keratinocytes. J Exp Med. 1994;179(4):1317–30.
50. Satsu H, Ishimoto Y, Nakano T, Mochizuki T, Iwanaga T, Shimizu M. Induction by activated macrophage-like THP-1 cells of apoptotic and necrotic cell death in intestinal epithelial Caco-2 monolayers via tumor necrosis factor-alpha. Exp Cell Res. 2006;312(19):3909–19.
51. Iwanaga T. The involvement of macrophages and lymphocytes in the apoptosis of enterocytes. Arch Histol Cytol. 1995;58(2):151–9.
52. Merger M, Viney JL, Borojevic R, Steele-Norwood D, Zhou P, Clark DA, Riddell R, Maric R, Podack ER, Croitoru K. Defining the roles of perforin, Fas/FasL, and tumour necrosis factor alpha in T cell induced mucosal damage in the mouse intestine. Gut. 2002;51(2):155–63.
53. Iwanaga T. Choujouhisaibou no Saibousi to sono Haijo (Apoptosis of intestinal epithelial cells and their disposal). Igaku no ayumi (J Clin Exp Med). 2008;225(6):507–10.
54. Huang FP, Platt N, Wykes M, Major JR, Powell TJ, Jenkins CD, MacPherson GG. A discrete subpopulation of dendritic cells transports apoptotic intestinal epithelial cells to T cell areas of mesenteric lymph nodes. J Exp Med. 2000;191(3):435–44.
55. Iida T, Onodera K, Nakase H. Role of autophagy in the pathogenesis of inflammatory bowel disease. World J Gastroenterol. 2017;23(11):1944–53.
56. Jostins L, Ripke S, Weersma RK, Duerr RH, McGovern DP, Hui KY, Lee JC, Schumm LP, Sharma Y, Anderson CA, Essers J, Mitrovic M, Ning K, Cleynen I, Theatre E, Spain SL, Raychaudhuri S, Goyette P, Wei Z, Abraham C, Achkar JP, Ahmad T, Amininejad L, Ananthakrishnan AN, Andersen V, Andrews JM, Baidoo L, Balschun T, Bampton PA, Bitton

A, Boucher G, Brand S, Büning C, Cohain A, Cichon S, D'Amato M, De Jong D, Devaney KL, Dubinsky M, Edwards C, Ellinghaus D, Ferguson LR, Franchimont D, Fransen K, Gearry R, Georges M, Gieger C, Glas J, Haritunians T, Hart A, Hawkey C, Hedl M, Hu X, Karlsen TH, Kupcinskas L, Kugathasan S, Latiano A, Laukens D, Lawrance IC, Lees CW, Louis E, Mahy G, Mansfield J, Morgan AR, Mowat C, Newman W, Palmieri O, Ponsioen CY, Potocnik U, Prescott NJ, Regueiro M, Rotter JI, Russell RK, Sanderson JD, Sans M, Satsangi J, Schreiber S, Simms LA, Sventoraityte J, Targan SR, Taylor KD, Tremelling M, Verspaget HW, De Vos M, Wijmenga C, Wilson DC, Winkelmann J, Xavier RJ, Zeissig S, Zhang B, Zhang CK, Zhao H, International IBD Genetics Consortium (IIBDGC), Silverberg MS, Annese V, Hakonarson H, Brant SR, Radford-Smith G, Mathew CG, Rioux JD, Schadt EE, Daly MJ, Franke A, Parkes M, Vermeire S, Barrett JC, Cho JH. Host-microbe interactions have shaped the genetic architecture of inflammatory bowel disease. Nature. 2012;491(7422):119–24.
57. Levine B, Mizushima N, Virgin HW. Autophagy in immunity and inflammation. Nature. 2011;469(7330):323–35.
58. Hooper KM, Barlow PG, Stevens C, Henderson P. Inflammatory bowel disease drugs: a focus on autophagy. J Crohns Colitis. 2017;11(1):118–27.
59. Shao B-Z, Han B-Z, Zeng Y-X, Su D-F, Liu C. The roles of macrophage autophagy in atherosclerosis. Acta Pharmacol Sin. 2016;37(2):150–6.
60. Liu NA, Shi Y, Zhuang S. Autophagy in chronic kidney diseases. Kidney Dis (Basel). 2016;2(1):37–45.
61. Gump JM, Thorburn A. Autophagy and apoptosis: what is the connection? Trends Cell Biol. 2011;21(7):387–92.
62. Canonico B, Cesarini E, Salucci S, Luchetti F, Falcieri E, Di Sario G, Palma F, Papa S. Defective autophagy, mitochondrial clearance and lipophagy in niemann-pick type B lymphocytes. PLoS One. 2016;11(10):e0165780.
63. Deretic V. Autophagy in leukocytes and other cells: mechanisms, subsystem organization, selectivity, and links to innate immunity. J Leukoc Biol. 2016;100(5):969–78.
64. Gao S, Sun D, Wang G, Zhang J, Jiang Y, Li G, Zhang K, Wang L, Huang J, Chen L. Growth inhibitory effect of paratocarpin E, a prenylated chalcone isolated from Euphorbia humifusa Wild., by induction of autophagy and apoptosis in human breast cancer cells. Bioorg Chem. 2016;69:121–8.
65. Li Y, Yu G, Yuan S, Tan C, Xie J, Ding Y, Lian P, Fu L, Hou Q, Xu B, Wang H. 14,15-Epoxyeicosatrienoic acid suppresses cigarette smoke condensate-induced inflammation in lung epithelial cells by inhibiting autophagy. Am J Physiol Lung Cell Mol Physiol. 2016;311(5):L970–80.
66. Wells JM, Rossi O, Meijerink M, Van Baarlen P. Epithelial crosstalk at the microbiota-mucosal interface. Proc Natl Acad Sci U S A. 2011;108(Suppl 1):4607–14.
67. Goto Y, Kiyono H. Epithelial barrier: an interface for the cross-communication between gut flora and immune system. Immunol Rev. 2012;245(1):147–63.
68. Moran AP, Gupta A, Joshi L. Sweet-talk: role of host glycosylation in bacterial pathogenesis of the gastrointestinal tract. Gut. 2011;60(10):1412–25.
69. Hansson GC. Role of mucus layers in gut infection and inflammation. Curr Opin Microbiol. 2012;15(1):57–62.
70. Cheng H, Leblond CP. Origin, differentiation and renewal of the four main epithelial cell types in the mouse small intestine. V. Unitarian Theory of the origin of the four epithelial cell types. Am J Anat. 1974;141(4):537–61.
71. Okumura R, Kurakawa T, Nakano T, Kayama H, Kinoshita M, Motooka D, Gotoh K, Kimura T, Kamiyama N, Kusu T, Ueda Y, Wu H, Iijima H, Barman S, Osawa H, Matsuno H, Nishimura J, Ohba Y, Nakamura S, Iida T, Yamamoto M, Umemoto E, Sano K, Takeda K. Lypd8 promotes the segregation of flagellated microbiota and colonic epithelia. Nature. 2016;532(7597):117–21.
72. Cadwell K, Patel KK, Maloney NS, Liu TC, Ng AC, Storer CE, Head RD, Xavier R, Stappenbeck TS, Virgin HW. Virus-plus-susceptibility gene interaction determines Crohn's disease gene *Atg16L1* phenotypes in intestine. Cell. 2010;141(7):1135–45.

73. Okamoto R, Watanabe M. Functional relevance of intestinal epithelial cells in inflammatory bowel disease. Nihon Rinsho Meneki Gakkai Kaishi (Jpn J Clin Immunol). 2016;39(6):522–7.
74. de Souza HSP, Fiocchi C. Immunopathogenesis of IBD: current state of the art. Nat Rev Gastroenterol Hepatol. 2016;13(1):13–27.
75. Inohara N, Ogura Y, Nuñez G. Nods: a family of cytosolic proteins that regulate the host response to pathogens. Curr Opin Microbiol. 2002;5(1):76–80.
76. Wehkamp J, Harder J, Weichenthal M, Schwab M, Schäffeler E, Schlee M, Herrlinger KR, Stallmach A, Noack F, Fritz P, Schröder JM, Bevins CL, Fellermann K, Stange EF. *NOD2* (CARD15) mutations in Crohn's disease are associated with diminished mucosal alpha-defensin expression. Gut. 2004;53(11):1658–64.

# Chapter 7
# Autophagic Regulation of Cardiomyocyte Survival and Heart Regeneration

**Toshihiko Aki, Kanako Noritake, Kana Unuma, and Koichi Uemura**

**Abstract** Autophagy is regarded as an essential cellular protective mechanism of cardiomyocytes against a panel of stresses such as myocardial infarction and ischemia–reperfusion. Autophagy-dependent protection against such stresses is especially important in cardiomyocytes, since adult cardiomyocytes are terminally differentiated cells and considered not to proliferate any more. However, this concept of adult cardiomyocyte nonproliferation has recently been challenged by many studies. Although the presence of cardiac stem cells in adult heart remains a subject of debate, there is ample evidence for the presence of cardiac progenitor cells that can differentiate into several heart-resident cells. Furthermore, adult cardiomyocytes can reenter the cell cycle and proliferate upon activation of YAP, a transcriptional coactivator downstream of the hippo pathway. In addition, cardiac cells in the epicardium can also transform into cardiofibroblasts, which contribute to tissue regeneration by filling damaged parts of tissue with themselves as well as extracellular matrix. This process seems to be executed through epithelial–mesenchymal transition (EMT). Autophagy has been supposed to participate in the maintenance of cardiomyocyte homeostasis not only by protecting the cells against stress, but also by facilitating regeneration. In this chapter, we discuss the possible roles of autophagy in protection as well as the promotion of regeneration of cardiomyocytes by regulating the hippo/YAP pathway and EMT.

## Abbreviations

| | |
|---|---|
| 8-OHdG | 8-Oxo-2′-deoxyguanosine |
| Atg | Autophagy-related |
| BH | Bcl-2 homologous |
| CMA | Chaperone-mediated autophagy |
| CTGF | Connective tissue growth factor |

T. Aki (✉) · K. Noritake · K. Unuma · K. Uemura
Department of Forensic Medicine, Tokyo Medical and Dental University, Tokyo, Japan
e-mail: aki.legm@tmd.ac.jp



K. Turksen (ed.), *Autophagy in Health and Disease*, Stem Cell Biology and Regenerative Medicine, https://doi.org/10.1007/978-3-319-98146-8_7

| | |
|---|---|
| EMT | Epithelial–mesenchymal transition |
| FIP200 | Focal adhesion kinase family interacting protein of 200 kDa |
| LAMP | Lysosome-associated membrane protein |
| LATS | Large tumor suppressor homolog |
| LC3 | Microtubles-associated protein light chain3 |
| MET | Mesenchymal–epithelial transition |
| Mst | Mammalian STE20-like protein kinase |
| RASSF1 | Ras association domain-containing protein 1 |
| ROS | Reactive oxygen species |
| Sav | Salvador homolog |
| TAZ | Transcriptional coactivator with PDZ-binding motif |
| TEAD | Transcriptiona enhancer-associated domain |
| ULK | Unc-51-like kinase |
| YAP | Yes-associated protein |

## 7.1 Proliferation of Mammalian Cardiomyocytes

Mammalian adult heart is composed of various cells such as cardiomyocytes, cardiofibroblasts, epicardial cells, and vascular endothelial and smooth muscle cells. In developed and industrialized nations, cardiovascular diseases are among the most frequent causes of human mortality. This is because the heart is an essential organ in vertebrates due to its critical role in the circulation of oxygen, nutrients, growth factors, and other essential molecules required by multicellular organisms. On the other hand, cardiomyocyte do not proliferate in the adult heart in mammals, suggesting the risk that cardiomyocyte death and resultant heart damage can lead directly to the death of an individual organism. During embryogenesis, cardiomyocytes are generated both through the differentiation of cardiac progenitor cells and the proliferation of existing cardiomyocytes [1–5]. In mammals, the primary structure of the heart is completed before birth, and after 1–2 weeks following birth, hypertrophy becomes the main contributor to the expansion of cardiac volume. In other words, cardiomyocytes exit the cell cycle soon after birth. Recently, a transcription factor called Meis1 has been suggested to be responsible for the arrest of the cardiomyocyte cell cycle after birth [6]. In mice, the expression of Meis1 is induced rapidly after birth and is responsible for the induction of several important cyclin-dependent-kinase inhibitors [7]. Indeed, it has also been shown that in the Meis1-knock out (KO) mice, cardiomyocytes continue to proliferate to repair injuries in the heart [6]. Another feature of mammalian adult cardiomyocytes is that substantial populations of cells are multinucleated, and many nuclei contain more than one set of chromosomes [8]. In contrast to mammals, zebrafish can regenerate cardiomyocytes even during adulthood [9, 10]. All of the cardiomyocytes in adult zebrafish are mononuclear and contain one set of chromosomes per nucleus, suggesting that adult cardiomyocytes still undergo proper cell cycle progression in zebrafish.

Because of the demand of blood circulation for survival, the heart is created during the early periods of embryogenesis. Although mammalian cardiomyocytes can proliferate during the fetal and neonatal periods, adult cardiomyocytes are terminally differentiated and ordinarily no longer proliferate. It has been shown in mice that cardiomyocytes can regenerate until 7 days after birth, but that this ability is lost between 7 and 21 days after birth [11]. This inability of adult cardiomyocyte to proliferate prevents the heart from regenerating following severe damage such as that caused by myocardial infarction. This lack of proliferative capability of cardiomyocytes is in sharp contrast to hepatocytes, which can regenerate to compensate for even ~70% loss of the whole liver even after the completion of development [12]. Skeletal muscle can also easily regenerate because of the existence of satellite cells, which possess stem cell-like properties and can undergo self-renewal as well as differentiation into myoblasts [13, 14]. Damaged cardiomyocytes are eliminated after their death and, in contrast to liver and skeletal muscle, damaged cardiac tissues have been believed to be replaced solely with cardiac fibroblasts, and not with new cardiomyocytes. Therefore, these scar tissues are detrimental to cardiac performance and lead eventually to heart failure.

Although it has long been believed that adult cardiomyocytes do not proliferate, recent research has shown that the turnover of adult cardiomyocyte occurs throughout life. Evidence for the renewal of adult cardiomyocytes in humans was shown by the use of carbon isotope analysis based on the incorporation of $^{14}C$ generated by the explosions of nuclear weapons during the Cold War era. Since 1963, when the test ban treaty for nuclear weapons was adopted, the amount of $^{14}C$ in the atmosphere has been declining exponentially, and the $^{14}C$ concentration in human tissues has also been declining. However, $^{14}C$ incorporated into the genome of postmitotic cells during the Cold War era should retain the concentration of $^{14}C$ at that time, since the half-life of $^{14}C$ is very long (5730 years). Based on this fact, Bergman et al. estimated the percentage of cardiomyocyte renewal as 1% in humans aged 25 and 0.45% at age 75 [15]. Other studies using isotope [16] as well as fluorescence [17] labeling methods confirmed the natural renewal of cardiomyocytes in mice, suggesting that, although rate is low, postmitotic mammalian adult cardiomyocytes can reenter the cell cycle and proliferate. Recent research has pointed out that oxidative DNA damage is one of the reasons for the inability of adult cardiomyocytes to proliferate [18]. Since cardiomyocytes have a large number of mitochondria due to the demand for high levels of ATP for actomyosin contraction, oxidative DNA damage, such as caused by the generation 8-OHdG, accumulates in the cells through mitochondrial ROS generation. Oxidized DNA accumulation triggers the DNA-damage response, which leads to mitotic arrest after birth. Indeed, it has been shown that the natural renewal of cardiomyocytes is accelerated under hypoxic conditions, confirming that mitochondrial oxygen-dependent respiration is involved in the mitotic arrest of cardiomyocytes [19].

It has been suggested that there are c-kit-positive cardiac progenitor/stem cells (CSCs) that could differentiate into not only cardiomyocytes, but also into vascular smooth muscle cells and endothelial cells in the heart [1, 20]. However, the contribution of CSCs to the repair of damaged heart is still controversial [21, 22]. In contrast to the as yet controversial role of these endogenous progenitor/stem cells in heart repair, the reprogramming of preresident cardiac fibroblasts into cardiomyocytes via cardiac transcription factors (GATA4, MEF2c, and Tbx5) is regarded as a promising method [23, 24].

Taken together, recent research has shown that adult cardiomyocyte renewal occurs, but that most cells remain outside of the cell cycle progression cycle. The postmitotic nature of adult cardiomyocytes has forced the cells to develop highly efficient processes to maintain homeostasis. Autophagy is one of these processes essential for cardiomyocyte survival in the face of stresses.

## 7.2 Process and Function of Autophagy

Autophagy is a lysosome-based bulk protein degradation system first pointed out by Christian de Duve, a Nobel laureate who discovered lysosomes, the cellular digestive organelle [25, 26]. The core machinery of autophagy was elucidated in budding yeast (*Saccharomyces cerevisiae*) by another Nobel laureate, Yoshinori Ohsumi. Extensive studies by Ohsumi's group discovered Atg (*Au*tophagy-related) proteins required for the execution of autophagy [27]. Up until now, at least 35 Atg genes have been reported in budding yeast. Although autophagy was discovered as a protein degradation system to recycle amino acids during periods of starvation [28, 29], it has been shown that autophagy involves the degradation of not only proteins but also lipids [30] and organelles [31]. Moreover, autophagy is involved in the elimination of bacteria [32], fungi [33], and viruses [34], suggesting that the machinery of autophagy is used to degrade not only intracellular substances but also extracellular xenobiotics.

There are at least three types of autophagy: macroautophagy, microautophagy, and chaperone-mediated autophagy (CMA) (Fig. 7.1). Microautophagy is a process during which cytoplasmic materials are engulfed by lysosomes and digested; however, the molecular mechanisms by which this process is executed in unicellular as well as multicellular organisms has been poorly understood [35, 36]. Like microautophagy, cytoplasmic materials are also delivered directly into lysosomes during chaperone-mediated autophagy (CMA). Various machineries have been identified as necessary for CMA [37]. First, hsc-70 binds to KFERQ-like amino acid sequences, which causes them to become sequestered to lysosomes [38]. LAMP2A, a splicing isoform generated from the LAMP2 gene, receives hsc-70 bounded proteins at the lysosomal membrane [39]. Specific proteins, including GAPDH, RNAse-A [40], and α-synuclein [41], have been suggested as CMA substrates. When it comes to cardiomyocytes, the type-2 ryanodine receptor has recently been shown to be a CMA substrate [42]. It has been shown that liver-specific CMA-KO mice develop hepatosteatosis, suggesting a role of CMA in liver homeostasis. However, the role of CMA in heart homeostasis remains to be discovered. Macroautophagy, as compared to microautophagy and CMA, is a dynamic membranous process [43, 44]. While cytoplasmic materials are delivered directly from the

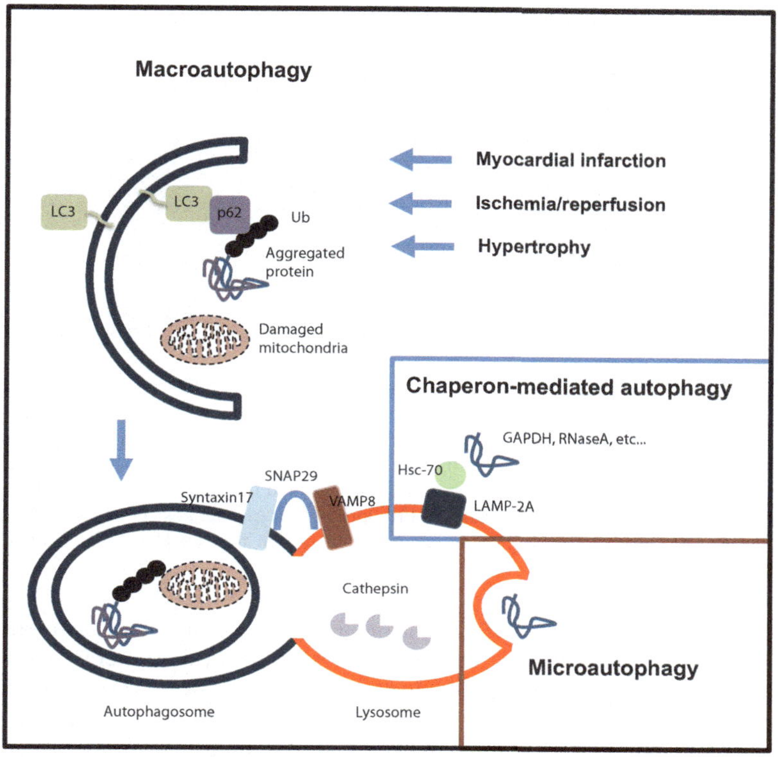


**Fig. 7.1** Schematic representation of autophagy process in cardiomyocytes

cytoplasm into lysosomes during microautophagy and CMA, during macroautophagy, materials are isolated by specified vesicles called autophagosomes. The first step in macroautophagy is the creation of the phogophore, which gradually elongates, and the closure of the elongated phagophore results in the formation of a double membrane structure (autophagosome) in which cytoplasmic materials are enclosed (Fig. 7.2). Autophagosomes fuse to lysosomes and the materials inside them are degraded by lysosomal hydrolyases. During cargo-specific autophagy, such as mitophagy, specified adaptors, such as p62, bind to LC3 on autophagosomes. Hereafter in this chapter, the term "autophagy" will be used to mean "macroautophagy."

The core machinery of autophagy involves the formation of the ULK (Unc51-like kinase, mammalian homolog of yeast Atg1) complex, which comprises ULK1/2, FIP200 (focal adhesion kinase family interacting protein of 200 kDa), Atg13, and Atg101. The activation of the ULK1/2 complex facilitates the formation of another complex, the beclin-1/classIII PI3K (phosphatidylinositol 3-kinase) complex [45], by the direct phosphorylation of beclin-1 at Ser-14. This complex generates PI3P (phosphatidylinositol 3-phosphate), which is required for the initiation of

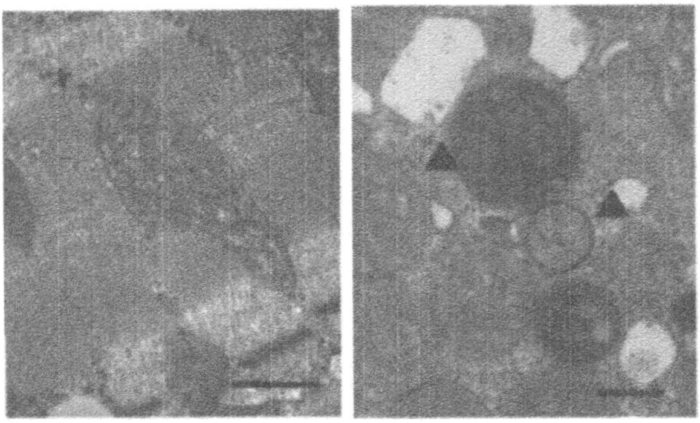

**Fig. 7.2** Transmission electron microscopic image of autophagy in rat heart. In normal heart (left panel), mitochondria are localized to the side of the sarcomere structure. In damaged heart (right panel, LPS-administered rat heart), almost all the mitochondria are swollen and many vacuoles are seen. Some mitochondria are surrounded by membranous structures (arrows), suggesting they will be eliminated through autophagy. Scale bars, 500 nm. Adopted from "Unuma K, Aki T, Funakoshi T, Yoshida K-i, Uemura K (2013) Cobalt Protoporphyrin Accelerates TFEB Activation and Lysosome Reformation during LPS Induced Septic Insults in the Rat Heart. PLoS ONE 8(2): e56526. doi:10.1371/journal.pone.0056526"

autophagosome formation. The expansion of the autophagosome membrane is mediated by two ubiquitin-like conjugation systems [46] that interact with each other. One ubiquitin-like conjugation system is the Atg12–Atg5 system. The ubiquitin-like molecule Atg-12 is activated sequentially by E1-like Atg7 and E2-like Atg10, and thereby conjugated to Atg5. Another ubiquitin-like system is the Atg8 (LC3)-PE system. LC3 is first cleaved by Atg4 to leave a glycine residue at its C-terminus [47]. Subsequently, the cleaved LC3 is activated by E1-like Atg7 and E2-like Atg3, finally conjugated to phosphatidylethanolamine (PE) via the C-terminal glycine residue [47]. The formation of Atg12–Atg5 has been shown to be a prerequisite for the LC3-PE system, suggesting that these two systems operate in parallel during the production of autophagosomes. The fusion between an autophagosome and lysosome is executed in a manner similar to other membrane fusion events, by SNARE proteins. Syntaxin17 has been shown to be the autophagosome-specific SNARE for this membrane fusion event [48]. Syntaxin 17 on the outer autophagosomal membrane recruits SNAP29 (synaptosomal-associated protein 29) and interacts with lysosomal VAMP8 (vesicle-associated membrane protein 8) to form a complex (Syntaxin 17-SNAP29-VAMP8 complex), which allows autophagome–lysosome fusion [48].

The most typical triggering event that induces autophagy is nutritional starvation. Mechanically, in mammals during starvation, decreased levels of amino acids and ATP are sensed by mTORC1 [mammalian target of rapamycin (mTOR) complex 1] and AMPK, respectively. Sufficient levels of amino acids render mTORC1 on lysosomes through a tetrametric complex of Rag GTPase [49]. mTOR on the lysosomal membrane phosphorylates several components of the ULK complex, thereby inhibiting the activity of this kinase complex to suppress autophagy [50].

On the other hand, cellular ATP levels are transmitted to the gamma subunit of AMPK, in which AMP binding to this subunit renders the kinase insensitive to upstream LKB1 kinase [51, 52]. Decreased ATP levels (increased AMP levels) activate AMPK, and this kinase in turn phosphorylates a component of mTORC1, raptor [53]. The phosphorylation of raptor inhibits mTORC1 activity, thereby inducing autophagy [54]. TFEB (transcription factor EB) has been supposed to act as a master transcription factor for autophagy as well as lysosome genesis [55, 56] (Fig. 7.3).

Alternative (macro) autophagy, which is independent of both Atg5 and Atg7, has recently been suggested [57]. The microscopic features of autophagy, such as the engulfment of cytoplasmic materials into double membranous autophagosomes, are also observed in alternative autophagy. However, the formation of LC3-II, a frequently used marker of macroautophagy, is not observed during alternative autophagy. Close inspection has shown that the membrane required for alternative autophagy should be derived from the Golgi apparatus [58], in contrast to conventional autophagy during which the autophagosomal membrane is derived mainly from ER. To date, only a few roles for alternative autophagy, such as the elimination of mitochondria during erythrogenesis [59], have been attributed both physiologically as well as pathophysiologically. The discovery of alternative autophagy points to the need to reconsider the role of autophagy in the homeostasis of the human body including heart, because many studies use Atg5- and Atg7-KO mice as model animals, or LC3-II as a marker, to show the involvement of autophagy. Indeed, it has been pointed out that Atg5-independent alternative autophagy is required for the generation of iPSC (induced pluripotent stem cells) through its requirement for eliminating mitochondria during cellular transformation into stem cells [60]. In addition, autophagy has been shown to promote longevity in *Drosophila* [61] and mice [62].

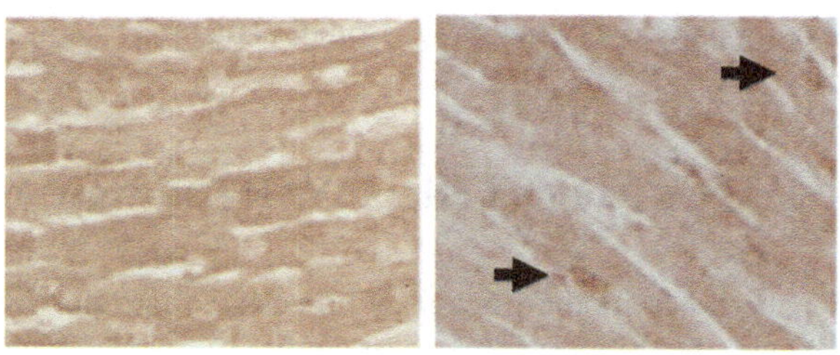

**Fig. 7.3** Activation of TFEB in damaged heart tissue. In normal heart (left panel), immunohistochemical analysis reveals the cytoplasmic localization of TFEB. In contrast, TFEB is found in the nucleus (arrows) in damaged heart tissue (right panel). Adopted from "Unuma K, Aki T, Funakoshi T, Yoshida K-i, Uemura K (2013) Cobalt Protoporphyrin Accelerates TFEB Activation and Lysosome Reformation during LPS Induced Septic Insults in the Rat Heart. PLoS ONE 8(2): e56526. doi:10.1371/journal.pone.0056526"

## 7.3 Autophagy and the Homeostasis of Mammalian Adult Cardiomyocytes

Since nutrient starvation is one of the main components of autophagy-inducing cardiac damage such as ischemia, it is quite natural that autophagy plays essential roles in maintaining cardiac homeostasis. Although the roles of microautophagy, CMA, and alternative autophagy in cardiomyocyte homeostasis have not been examined so far, there is ample evidence for the essential role of autophagy in cardiomyocyte homeostasis. Autophagy has been shown to be involved in the protection of cardiomyocytes from injuries caused by aortic occlusion, myocardial infarction, and subsequent hemodynamic stresses [63, 64]. The essential roles of autophagy in cardiomyocytes were observed for the first time during the analysis of the mouse deleted LAMP2 gene. In hearts of LAMP2-KO mice, numerous autophagic vacuoles were observed, suggesting that cardiomyocytes undergo autophagy under basal conditions: autophagosome/autolysosomes should accumulate in the heart when lysosomal function is severely impaired due to the lack of LAMP2 [65]. This phenotype resembles that observed in Danon disease, a lysosomal storage disorder associated with cardiomyopathy. The same cardiac phenotype is also observed in humans harboring the mutation for familial-type Danon disease in the LAMP2 gene [66]. Further, it has been shown that cardiomyocyte-specific Atg5-KO results in no outcome under healthy conditions, but leads to decreased heart performance with the concomitant loss of cardiomyocytes by apoptosis upon transaortic constriction [63]. Indeed, studies using autophagy inhibitors have shown that autophagy participates in protecting cells during reperfusion injuries after ischemia, while autophagy itself is detrimental to ischemic heart injuries [64]. This observation, together with the accumulation of autophagic vacuoles in the heart of LAMP2-KO mouse, shows that incomplete autophagy does not fulfill the cytoprotective role and is rather detrimental to cells. Since lysosomes consume a lot of cellular ATP to acidify its inside milieu though the operation of proton-ATPase, the decline in ATP during ischemia should also result in a decline in lysosomal function. Incomplete autophagy or autophagy failure might occur when lysosomal function is severely impaired and/or when excessive autophagy overwhelms lysosomal capacity.

## 7.4 Process and Function of Hippo/YAP Pathway

The hippo/YAP pathway is another important process for cardiomyocyte homeostasis. Like autophagy, the hippo pathway, including its downstream target Yes-associated protein (YAP), is a highly conserved process among higher organisms and *Drosophila*. The hippo pathway molecules were first identified as genes whose mutation causes tissue overgrowth in *Drosophila*. During the 2000s, *Drosophila* Hippo, Warts, and Yorkie were shown to correspond to mammalian Ste20 like kinase 1/2 (Mst1/2), large tumor suppressor 1/2 (LATS 1/2), and YAP, respectively

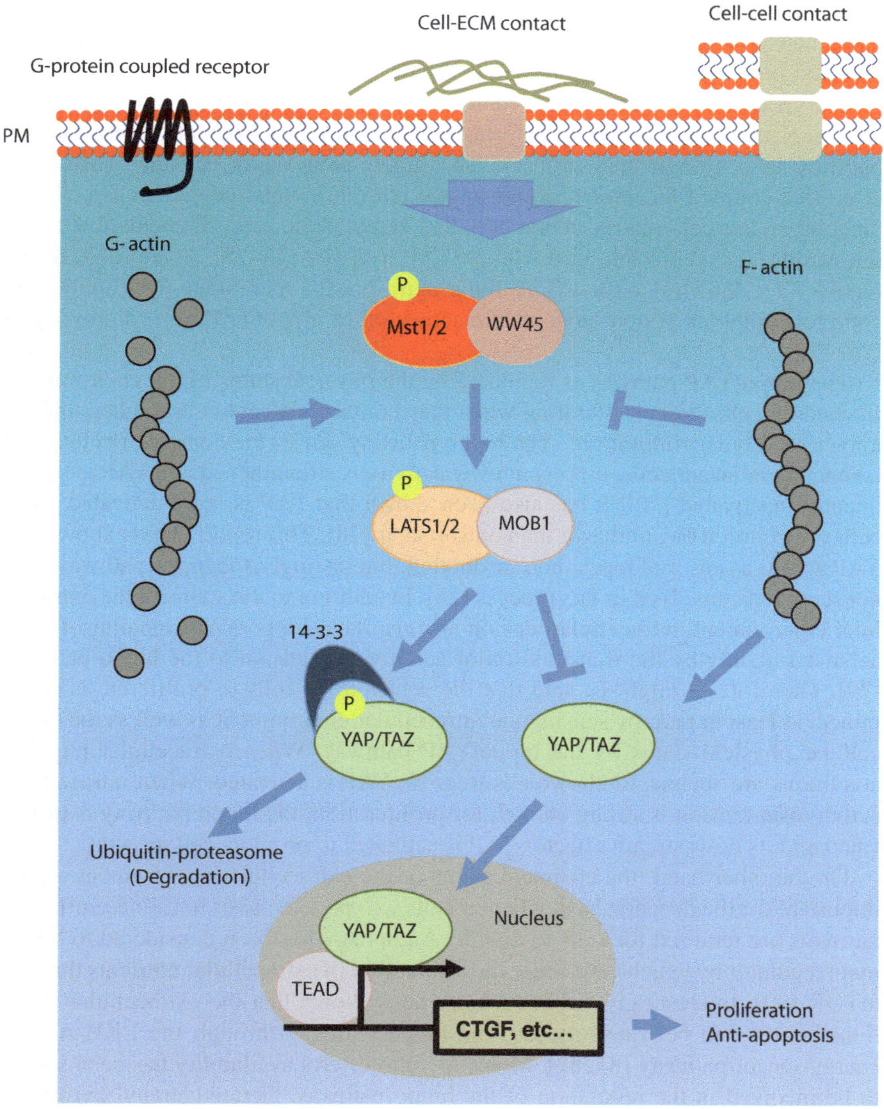


**Fig. 7.4** Schematic representation of the hippo/YAP pathway

[67–69] (Fig. 7.4). The core pathway of hippo signaling is composed of Mst and LATS, and their associated molecules WW45 (45 kDa WW domain protein) and MOB1 (mps one binder 1), respectively. In mammals, LATS phosphorylates YAP, and phosphorylated YAP is bound to 14-3-3 and retained in the cytoplasm [70]. Upon inactivation of the hippo pathway, YAP is de-phosphorylated and enters the nucleus [71] where it acts as a transcriptional coactivator of the TEAD transcriptional activator complex [72]. Target genes of the YAP-TEAD transcriptional activator complex include CTGF (connective tissue growth factor), which is involved in the growth of connective and other tissues [72]. These genes facilitate cell survival and proliferation. In contrast to plasma membrane receptor-mediated signaling pathways, such as growth factors and cytokines, regulation of the hippo pathway is a complicated process. The hippo pathway is activated through a G-protein-coupled receptor residing on the plasma membrane, adherens cell–cell junction and/or cell–matrix interaction, and cytoplasmic actin filaments that transmit cellular tension to inhibit the hippo/YAP pathway [73–75]. In addition, recent reports have shown that the extracellular agrin-plasma membrane dystrophin–glycoprotein complex system governs the activation of hippo/YAP in cardiomyocytes [76, 77].

The hippo/YAP pathway is regulated by the physical status of the environment. In general, cells stop proliferating when they contact adjacent cells, a phenomenon known as "contact inhibition." The hippo pathway senses the density of cells: after contacting adjacent cells, hippo pathway activity is strengthened and YAP is subsequently inactivated [70]. It has also been shown that YAP is more activated when cells are cultured on solid agar than on soft agar [78]. This report clearly shows that YAP acts as a sensor of mechanotransduction. Interestingly, the hippo pathway does not seem to be involved in this process [78]. In addition to the status of the extracellular environment, intracellular tension also regulates hippo/YAP signaling. This is achieved mainly by the transmission of actomyosin tension to the hippo pathway [79]. Therefore, it might be said that the decision for cells to proliferate is determined, at least in part, by sensing the intracellular environment as well as the extracellular physical status by the hippo/YAP pathway. When extracellular physical conditions are such as to allow cells to grow, YAP is activated. When intracellular actomyosin tension is strong enough for proliferation, the hippo pathway is inactivated and its downstream effector YAP is activated to promote cell growth.

On the other hand, the chemical status of the extracellular environment is also important for the decision as to whether cells will proliferate or not, since sufficient nutrients are required for cells to double. Autophagy has been considered to be the main regulator by which cells sense the availability of extracellular nutrients through mTOR [80]. Interestingly, it has recently been shown that the extracellular nutritional status can be transmitted to the hippo pathway through the LKB-AMPK energy sensor pathway [81, 82]. Moreover, cholesterol availability has been shown to be involved in the regulation of the hippo pathway. Geranylgeranyl pyrophosphate, one of the intermediate metabolites in the cholesterol synthesis pathway (mevalonate pathway) acts as a repressor of the hippo pathway through direct geranylgeranylation of Rho-kinase [83]. Since the availability of cholesterol as a source of cellular membranes is also required for cell proliferation, this connection between the cellular lipid synthesis pathway and growth regulating pathway represents an elegant example of a cellular fine-tuning system.

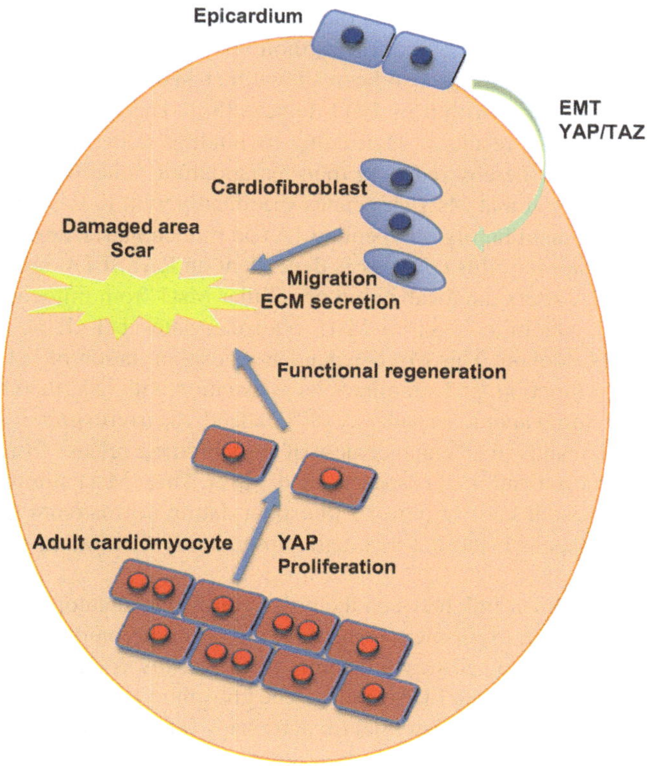


**Fig. 7.5** Schematic representation of the hippo/YAP pathway and EMT during cardiomyocyte regeneration

## 7.5 Roles of Hippo/YAP Pathway in Cardiomyocyte Proliferation

Recently, the emerging role of the proliferation-suppressing hippo-pathway and its downstream target, proliferation-stimulating Yes-associated protein (YAP), in cardiomyocyte proliferation has been extensively described (Fig. 7.5): it has been shown that gene transfer of YAP into adult cardiomyocytes gives them the ability to proliferate [84]. Consistently, suppression of the hippo pathway causes cardiomyocytes to proliferate through the nuclear translocation of YAP [85]. It has also been shown that the hippo pathway is involved in the regulation of autophagy. A direct link between the hippo/YAP pathway and autophagy has been suggested by several studies in various tissues including cardiac tissues. In a mouse model of cardiac ischemia created by coronary artery ligation, the forced expression of dominant-negative Mst1, a hippo pathway molecule, in cardiomyocytes enhanced autophagy and reduced the formation of aggresomes, including autophagy substrate p62 [86]. In mice harboring the heterozygous deletion of beclin1, the effects of Mst1 deficiency on autophagy as well as cardiac performance were canceled, suggesting that Mst1 exerts its negative effect on cardiomyocytes by regulating beclin1 and autophagy [86]. Indeed, it has also been shown that Mst1 directly phosphorylates beclin1 at residue $Thr^{108}$ within its BH3-domain [86]. The Mst1-dependent phosphorylation of beclin1 results in facilitating its binding to the antiapoptotic bcl-2 and bcl-xL proteins, thereby reducing their antiapoptotic function [86]. The same group also pointed out that Mst1 associates specifically with K-Ras and RASSF1A (ras-associated domain family polypeptide 1A) on mitochondria, and that the mitochondrial localization of this complex is dependent on RASSF1A, since the depletion of RASSF1A alters the localization of cellular Mst1 from mitochondria to the cytoplasm [87]. On mitochondria, Mst1 phosphorylates bcl-xL at $Ser^{14}$ in the N-terminal BH4 domain. This Mst1-dependent phosphorylation of bcl-xL impairs its antiapoptotic function, and stimulates its association with bax, thereby stimulating the mitochondrial apoptosis pathway [87]. Indeed, the overexpression of Mst1 in cardiomyocytes results in an enhancement of cytochrome c release from mitochondria as well as the activation of caspases 9 and 3 [87]. Thus, Mst1 stimulates cardiomyocyte apoptosis, at least in part, by phosphorylating downstream targets, bcl-2 family proteins (beclin1 and bcl-xL), and subsequently facilitating apoptosis and inhibiting autophagy.

Evidence for a direct link between the hippo pathway and autophagy machinery has also come from an experiment using both mouse embryonic fibroblasts and *C. elegans* [88]. Core hippo pathway molecules in *C. elegans*, STK3/4 (Mst2/1), have been shown to phosphorylate LC3 directly. The phosphorylation of LC3 at $Thr^{50}$ by STK3/4 has been shown to be essential for autophagy [88]. The depletion of STK3/4 or the replacement of $Thr^{50}$ in LC3 with a nonphosphorylatable amino acid results in the accumulation of autophagosomes, suggesting an impairment of fusion between autophagosomes and lysosomes [88]. Thus, the STK3/4-dependent phosphorylation of LC3 is essential for the creation of autolysosomes, at least in some types of mammalian cells.

## 7.6 Epithelial to Mesenchymal Transition (EMT) and Heart Regeneration

Although cardiomyocytes can proliferate even during adulthood, such proliferation is rare event in adults. Therefore, the cells/tissues lost due to cardiac injuries are ordinarily repaired by replacement with extracellular matrix (ECM) (Fig. 7.5). ECM synthesized by fibroblasts is important to fill the injured areas of tissues for regeneration, but excessive amounts of ECM impair the function of the tissue, something known as fibrosis. In the heart, fibroblasts (cardiofibroblasts) are generated from epicardium through epithelial to mesenchymal transition (EMT). EMT is a process by which epithelial cells transform their phenotype from epithelial to the phenotype of mesenchymal cells, such as smooth muscle cells and fibroblasts.

Epicardium is the tissue from which fibroblasts, vascular smooth muscle as well as endothelial cells are derived during the development of the heart. Loss of YAP/TAZ (transcriptional coactivator with a PDZ-binding motif, another mammalian homolog of *Drosophila* Yokie) in epicardium has been shown to affect vascular formation from epicardium during embryonic heart development [89]. YAP and TAZ are activated in the embryonic epicardium and direct the transformation of epicardial cells into cardiofibroblasts or vascular cells through EMT, suggesting a relationship between hippo/YAP signaling and EMT [89]. Interestingly, epicardial YAP/TAZ is also involved in the regeneration of injured myocardial tissue in adult heart, suggesting a role of YAP/TAZ and epicardium in the regeneration of the myocardium damaged by stresses [90].

## 7.7 Conclusion and Perspectives

Many studies have shown that autophagy is essential for cardiomyocyte survival against a panel of stresses including coronary occlusion, ischemic injury, and hypertrophy. Similarly, hippo/YAP has been shown to be necessary as well as sufficient for the regeneration of cardiomyocytes. Nevertheless, there are only a few reports describing the roles of autophagy in cardiomyocyte regeneration as well as hippo/YAP pathway regulation. Since autophagy is a catabolic rather than anabolic process, autophagy should not help cell proliferation directly. Indeed, it is generally accepted that autophagy is not advantageous for the late stages of tumorigenesis, during which immortalized cells proliferate extensively using the rich environmental materials provided by newly created microvessels. In contrast, autophagy is often needed during the early stages of tumorigenesis when cells need to survive under nutrient-poor conditions. Therefore, during the process of cardiomyocyte regeneration in response to heart-damaging insults, autophagy might be favored during the earlier periods when cells must survive their life-threatening stresses. On the other hand, autophagy might not be favored during the latter half of the process, since cell proliferation is an extensively anabolic process. However, heart regeneration is complicated, and extensive research is required to elucidate the precise role of autophagy in heart regeneration.

**Acknowledgments** This work was supported in part by MEXT KAKENHI Grant Number JP25460862. The authors declare no conflicts of interest.

## References

1. Leri A, Kajstura J, Anversa P. Role of cardiac stem cells in cardiac pathophysiology: a paradigm shift in human myocardial biology. Circ Res. 2011;109(8):941–61.
2. Sahara M, Santoro F, Chien KR. Programming and reprogramming a human heart cell. EMBO J. 2015;34(6):710–38.

3. Foglia MJ, Poss KD. Building and re-building the heart by cardiomyocyte proliferation. Development. 2016;143(5):729–40.
4. Uygur A, Lee RT. Mechanisms of cardiac regeneration. Dev Cell. 2016;36(4):362–74.
5. Tzahor E, Poss KD. Cardiac regeneration strategies: staying young at heart. Science. 2017;356(6342):1035–9.
6. Mahmoud AI, Kocabas F, Muralidhar SA, Kimura W, Koura AS, Thet S, Porrello ER, Sadek HA. Meis1 regulates postnatal cardiomyocyte cell cycle arrest. Nature. 2013;497(7448):249–53.
7. Ponnusamy M, Li PF, Wang K. Understanding cardiomyocyte proliferation: an insight into cell cycle activity. Cell Mol Life Sci. 2017;74(6):1019–34.
8. Laflamme MA, Murry CE. Heart regeneration. Nature. 2011;473(7347):326–35.
9. Poss KD, Wilson LG, Keating MT. Heart regeneration in zebrafish. Science. 2002;298(5601):2188–90.
10. Gonzalez-Rosa JM, Burns CE, Burns CG. Zebrafish heart regeneration: 15 years of discoveries. Regeneration (Oxf). 2017;4(3):105–23.
11. Porrello ER, Mahmoud AI, Simpson E, Hill JA, Richardson JA, Olson EN, Sadek HA. Transient regenerative potential of the neonatal mouse heart. Science. 2011;331(6020):1078–80.
12. Michalopoulos GK. Liver regeneration. J Cell Physiol. 2007;213(2):286–300.
13. Relaix F, Zammit PS. Satellite cells are essential for skeletal muscle regeneration: the cell on the edge returns centre stage. Development. 2012;139(16):2845–56.
14. Mauro A. Satellite cell of skeletal muscle fibers. J Biophys Biochem Cytol. 1961;9:493–5.
15. Bergmann O, Bhardwaj RD, Bernard S, Zdunek S, Barnabe-Heider F, Walsh S, Zupicich J, Alkass K, Buchholz BA, Druid H, Jovinge S, Frisen J. Evidence for cardiomyocyte renewal in humans. Science. 2009;324(5923):98–102.
16. Senyo SE, Steinhauser ML, Pizzimenti CL, Yang VK, Cai L, Wang M, Wu TD, Guerquin-Kern JL, Lechene CP, Lee RT. Mammalian heart renewal by pre-existing cardiomyocytes. Nature. 2013;493(7432):433–6.
17. Ali SR, Hippenmeyer S, Saadat LV, Luo L, Weissman IL, Ardehali R. Existing cardiomyocytes generate cardiomyocytes at a low rate after birth in mice. Proc Natl Acad Sci U S A. 2014;111(24):8850–5.
18. Puente BN, Kimura W, Muralidhar SA, Moon J, Amatruda JF, Phelps KL, Grinsfelder D, Rothermel BA, Chen R, Garcia JA, Santos CX, Thet S, Mori E, Kinter MT, Rindler PM, Zacchigna S, Mukherjee S, Chen DJ, Mahmoud AI, Giacca M, Rabinovitch PS, Aroumougame A, Shah AM, Szweda LI, Sadek HA. The oxygen-rich postnatal environment induces cardiomyocyte cell-cycle arrest through DNA damage response. Cell. 2014;157(3):565–79.
19. Nakada Y, Canseco DC, Thet S, Abdisalaam S, Asaithamby A, Santos CX, Shah AM, Zhang H, Faber JE, Kinter MT, Szweda LI, Xing C, Hu Z, Deberardinis RJ, Schiattarella G, Hill JA, Oz O, Lu Z, Zhang CC, Kimura W, Sadek HA. Hypoxia induces heart regeneration in adult mice. Nature. 2017;541(7636):222–7.
20. Beltrami AP, Barlucchi L, Torella D, Baker M, Limana F, Chimenti S, Kasahara H, Rota M, Musso E, Urbanek K, Leri A, Kajstura J, Nadal-Ginard B, Anversa P. Adult cardiac stem cells are multipotent and support myocardial regeneration. Cell. 2003;114(6):763–76.
21. van Berlo JH, Kanisicak O, Maillet M, Vagnozzi RJ, Karch J, Lin SC, Middleton RC, Marban E, Molkentin JD. c-kit+ cells minimally contribute cardiomyocytes to the heart. Nature. 2014;509(7500):337–41.
22. Ellison GM, Vicinanza C, Smith AJ, Aquila I, Leone A, Waring CD, Henning BJ, Stirparo GG, Papait R, Scarfo M, Agosti V, Viglietto G, Condorelli G, Indolfi C, Ottolenghi S, Torella D, Nadal-Ginard B. Adult c-kit(pos) cardiac stem cells are necessary and sufficient for functional cardiac regeneration and repair. Cell. 2013;154(4):827–42.
23. Ieda M, Fu JD, Delgado-Olguin P, Vedantham V, Hayashi Y, Bruneau BG, Srivastava D. Direct reprogramming of fibroblasts into functional cardiomyocytes by defined factors. Cell. 2010;142(3):375–86.
24. Sadahiro T, Yamanaka S, Ieda M. Direct cardiac reprogramming: progress and challenges in basic biology and clinical applications. Circ Res. 2015;116(8):1378–91.

25. Klionsky DJ. Autophagy revisited: a conversation with Christian de Duve. Autophagy. 2008;4(6):740–3.
26. Yin XM, Ding WX, Gao W. Autophagy in the liver. Hepatology. 2008;47(5):1773–85.
27. Tsukada M, Ohsumi Y. Isolation and characterization of autophagy-defective mutants of Saccharomyces cerevisiae. FEBS Lett. 1993;333(1-2):169–74.
28. Takeshige K, Baba M, Tsuboi S, Noda T, Ohsumi Y. Autophagy in yeast demonstrated with proteinase-deficient mutants and conditions for its induction. J Cell Biol. 1992;119(2):301–11.
29. Baba M, Takeshige K, Baba N, Ohsumi Y. Ultrastructural analysis of the autophagic process in yeast: detection of autophagosomes and their characterization. J Cell Biol. 1994;124(6):903–13.
30. Singh R, Kaushik S, Wang Y, Xiang Y, Novak I, Komatsu M, Tanaka K, Cuervo AM, Czaja MJ. Autophagy regulates lipid metabolism. Nature. 2009;458(7242):1131–5.
31. Anding AL, Baehrecke EH. Cleaning house: selective autophagy of organelles. Dev Cell. 2017;41(1):10–22.
32. Saitoh T, Akira S. Regulation of innate immune responses by autophagy-related proteins. J Cell Biol. 2010;189(6):925–35.
33. Kanayama M, Shinohara ML. Roles of autophagy and autophagy-related proteins in antifungal immunity. Front Immunol. 2016;7:47.
34. Dong X, Levine B. Autophagy and viruses: adversaries or allies? J Innate Immun. 2013;5(5):480–93.
35. Li WW, Li J, Bao JK. Microautophagy: lesser-known self-eating. Cell Mol Life Sci. 2012;69(7):1125–36.
36. Mijaljica D, Prescott M, Devenish RJ. Microautophagy in mammalian cells: revisiting a 40-year-old conundrum. Autophagy. 2011;7(7):673–82.
37. Orenstein SJ, Cuervo AM. Chaperone-mediated autophagy: molecular mechanisms and physiological relevance. Semin Cell Dev Biol. 2010;21(7):719–26.
38. Chiang HL, Terlecky SR, Plant CP, Dice JF. A role for a 70-kilodalton heat shock protein in lysosomal degradation of intracellular proteins. Science. 1989;246(4928):382–5.
39. Cuervo AM, Dice JF. A receptor for the selective uptake and degradation of proteins by lysosomes. Science. 1996;273(5274):501–3.
40. Cuervo AM, Terlecky SR, Dice JF, Knecht E. Selective binding and uptake of ribonuclease A and glyceraldehyde-3-phosphate dehydrogenase by isolated rat liver lysosomes. J Biol Chem. 1994;269(42):26374–80.
41. Cuervo AM, Stefanis L, Fredenburg R, Lansbury PT, Sulzer D. Impaired degradation of mutant alpha-synuclein by chaperone-mediated autophagy. Science. 2004;305(5688):1292–5.
42. Pedrozo Z, Torrealba N, Fernandez C, Gatica D, Toro B, Quiroga C, Rodriguez AE, Sanchez G, Gillette TG, Hill JA, Donoso P, Lavandero S. Cardiomyocyte ryanodine receptor degradation by chaperone-mediated autophagy. Cardiovasc Res. 2013;98(2):277–85.
43. He C, Klionsky DJ. Regulation mechanisms and signaling pathways of autophagy. Annu Rev Genet. 2009;43:67–93.
44. Mizushima N, Yoshimori T, Ohsumi Y. The role of Atg proteins in autophagosome formation. Annu Rev Cell Dev Biol. 2011;27:107–32.
45. Kihara A, Kabeya Y, Ohsumi Y, Yoshimori T. Beclin-phosphatidylinositol 3-kinase complex functions at the trans-Golgi network. EMBO Rep. 2001;2(4):330–5.
46. Ohsumi Y. Molecular dissection of autophagy: two ubiquitin-like systems. Nat Rev Mol Cell Biol. 2001;2(3):211–6.
47. Kabeya Y, Mizushima N, Ueno T, Yamamoto A, Kirisako T, Noda T, Kominami E, Ohsumi Y, Yoshimori T. LC3, a mammalian homologue of yeast Apg8p, is localized in autophagosome membranes after processing. EMBO J. 2000;19(21):5720–8.
48. Itakura E, Kishi-Itakura C, Mizushima N. The hairpin-type tail-anchored SNARE syntaxin 17 targets to autophagosomes for fusion with endosomes/lysosomes. Cell. 2012;151(6):1256–69.
49. Sancak Y, Bar-Peled L, Zoncu R, Markhard AL, Nada S, Sabatini DM. Ragulator-Rag complex targets mTORC1 to the lysosomal surface and is necessary for its activation by amino acids. Cell. 2010;141(2):290–303.

50. Kim J, Kundu M, Viollet B, Guan KL. AMPK and mTOR regulate autophagy through direct phosphorylation of Ulk1. Nat Cell Biol. 2011;13(2):132–41.
51. Kahn BB, Alquier T, Carling D, Hardie DG. AMP-activated protein kinase: ancient energy gauge provides clues to modern understanding of metabolism. Cell Metab. 2005;1(1):15–25.
52. Shaw RJ. LKB1 and AMP-activated protein kinase control of mTOR signalling and growth. Acta Physiol (Oxf). 2009;196(1):65–80.
53. Gwinn DM, Shackelford DB, Egan DF, Mihaylova MM, Mery A, Vasquez DS, Turk BE, Shaw RJ. AMPK phosphorylation of raptor mediates a metabolic checkpoint. Mol Cell. 2008;30(2):214–26.
54. Dunlop EA, Tee AR. The kinase triad, AMPK, mTORC1 and ULK1, maintains energy and nutrient homoeostasis. Biochem Soc Trans. 2013;41(4):939–43.
55. Sardiello M, Palmieri M, di Ronza A, Medina DL, Valenza M, Gennarino VA, Di Malta C, Donaudy F, Embrione V, Polishchuk RS, Banfi S, Parenti G, Cattaneo E, Ballabio A. A gene network regulating lysosomal biogenesis and function. Science. 2009;325(5939):473–7.
56. Settembre C, Di Malta C, Polito VA, Garcia Arencibia M, Vetrini F, Erdin S, Erdin SU, Huynh T, Medina D, Colella P, Sardiello M, Rubinsztein DC, Ballabio A. TFEB links autophagy to lysosomal biogenesis. Science. 2011;332(6036):1429–33.
57. Nishida Y, Arakawa S, Fujitani K, Yamaguchi H, Mizuta T, Kanaseki T, Komatsu M, Otsu K, Tsujimoto Y, Shimizu S. Discovery of Atg5/Atg7-independent alternative macroautophagy. Nature. 2009;461(7264):654–8.
58. Yamaguchi H, Arakawa S, Kanaseki T, Miyatsuka T, Fujitani Y, Watada H, Tsujimoto Y, Shimizu S. Golgi membrane-associated degradation pathway in yeast and mammals. EMBO J. 2016;35(18):1991–2007.
59. Honda S, Arakawa S, Nishida Y, Yamaguchi H, Ishii E, Shimizu S. Ulk1-mediated Atg5-independent macroautophagy mediates elimination of mitochondria from embryonic reticulocytes. Nat Commun. 2014;5:4004.
60. Ma T, Li J, Xu Y, Yu C, Xu T, Wang H, Liu K, Cao N, Nie BM, Zhu SY, Xu S, Li K, Wei WG, Wu Y, Guan KL, Ding S. Atg5-independent autophagy regulates mitochondrial clearance and is essential for iPSC reprogramming. Nat Cell Biol. 2015;17(11):1379–87.
61. Simonsen A, Cumming RC, Brech A, Isakson P, Schubert DR, Finley KD. Promoting basal levels of autophagy in the nervous system enhances longevity and oxidant resistance in adult Drosophila. Autophagy. 2008;4(2):176–84.
62. Pyo JO, Yoo SM, Ahn HH, Nah J, Hong SH, Kam TI, Jung S, Jung YK. Overexpression of Atg5 in mice activates autophagy and extends lifespan. Nat Commun. 2013;4:2300.
63. Nakai A, Yamaguchi O, Takeda T, Higuchi Y, Hikoso S, Taniike M, Omiya S, Mizote I, Matsumura Y, Asahi M, Nishida K, Hori M, Mizushima N, Otsu K. The role of autophagy in cardiomyocytes in the basal state and in response to hemodynamic stress. Nat Med. 2007;13(5):619–24.
64. Matsui Y, Takagi H, Qu X, Abdellatif M, Sakoda H, Asano T, Levine B, Sadoshima J. Distinct roles of autophagy in the heart during ischemia and reperfusion: roles of AMP-activated protein kinase and Beclin 1 in mediating autophagy. Circ Res. 2007;100(6):914–22.
65. Tanaka Y, Guhde G, Suter A, Eskelinen EL, Hartmann D, Lullmann-Rauch R, Janssen PM, Blanz J, von Figura K, Saftig P. Accumulation of autophagic vacuoles and cardiomyopathy in LAMP-2-deficient mice. Nature. 2000;406(6798):902–6.
66. Nishino I, Fu J, Tanji K, Yamada T, Shimojo S, Koori T, Mora M, Riggs JE, Oh SJ, Koga Y, Sue CM, Yamamoto A, Murakami N, Shanske S, Byrne E, Bonilla E, Nonaka I, DiMauro S, Hirano M. Primary LAMP-2 deficiency causes X-linked vacuolar cardiomyopathy and myopathy (Danon disease). Nature. 2000;406(6798):906–10.
67. Wu S, Huang J, Dong J, Pan D. hippo encodes a Ste-20 family protein kinase that restricts cell proliferation and promotes apoptosis in conjunction with salvador and warts. Cell. 2003;114(4):445–56.
68. Harvey KF, Pfleger CM, Hariharan IK. The Drosophila Mst ortholog, hippo, restricts growth and cell proliferation and promotes apoptosis. Cell. 2003;114(4):457–67.

69. Huang J, Wu S, Barrera J, Matthews K, Pan D. The Hippo signaling pathway coordinately regulates cell proliferation and apoptosis by inactivating Yorkie, the Drosophila Homolog of YAP. Cell. 2005;122(3):421–34.
70. Zhao B, Wei X, Li W, Udan RS, Yang Q, Kim J, Xie J, Ikenoue T, Yu J, Li L, Zheng P, Ye K, Chinnaiyan A, Halder G, Lai ZC, Guan KL. Inactivation of YAP oncoprotein by the Hippo pathway is involved in cell contact inhibition and tissue growth control. Genes Dev. 2007;21(21):2747–61.
71. Yagi R, Chen LF, Shigesada K, Murakami Y, Ito Y. A WW domain-containing yes-associated protein (YAP) is a novel transcriptional co-activator. EMBO J. 1999;18(9):2551–62.
72. Zhao B, Ye X, Yu J, Li L, Li W, Li S, Lin JD, Wang CY, Chinnaiyan AM, Lai ZC, Guan KL. TEAD mediates YAP-dependent gene induction and growth control. Genes Dev. 2008;22(14):1962–71.
73. Halder G, Dupont S, Piccolo S. Transduction of mechanical and cytoskeletal cues by YAP and TAZ. Nat Rev Mol Cell Biol. 2012;13(9):591–600.
74. Yu FX, Zhao B, Panupinthu N, Jewell JL, Lian I, Wang LH, Zhao J, Yuan H, Tumaneng K, Li H, Fu XD, Mills GB, Guan KL. Regulation of the hippo-YAP pathway by G-protein-coupled receptor signaling. Cell. 2012;150(4):780–91.
75. Panciera T, Azzolin L, Cordenonsi M, Piccolo S. Mechanobiology of YAP and TAZ in physiology and disease. Nat Rev Mol Cell Biol. 2017;18:758.
76. Bassat E, Mutlak YE, Genzelinakh A, Shadrin IY, Baruch Umansky K, Yifa O, Kain D, Rajchman D, Leach J, Riabov Bassat D, Udi Y, Sarig R, Sagi I, Martin JF, Bursac N, Cohen S, Tzahor E. The extracellular matrix protein agrin promotes heart regeneration in mice. Nature. 2017;547(7662):179–84.
77. Morikawa Y, Heallen T, Leach J, Xiao Y, Martin JF. Dystrophin-glycoprotein complex sequesters Yap to inhibit cardiomyocyte proliferation. Nature. 2017;547(7662):227–31.
78. Dupont S, Morsut L, Aragona M, Enzo E, Giulitti S, Cordenonsi M, Zanconato F, Le Digabel J, Forcato M, Bicciato S, Elvassore N, Piccolo S. Role of YAP/TAZ in mechanotransduction. Nature. 2011;474(7350):179–83.
79. Rauskolb C, Sun S, Sun G, Pan Y, Irvine KD. Cytoskeletal tension inhibits Hippo signaling through an Ajuba-Warts complex. Cell. 2014;158(1):143–56.
80. Wolfson RL, Sabatini DM. The dawn of the age of amino acid sensors for the mTORC1 pathway. Cell Metab. 2017;26(2):301–9.
81. Wang W, Xiao ZD, Li X, Aziz KE, Gan B, Johnson RL, Chen J. AMPK modulates Hippo pathway activity to regulate energy homeostasis. Nat Cell Biol. 2015;17(4):490–9.
82. Mo JS, Meng Z, Kim YC, Park HW, Hansen CG, Kim S, Lim DS, Guan KL. Cellular energy stress induces AMPK-mediated regulation of YAP and the Hippo pathway. Nat Cell Biol. 2015;17(4):500–10.
83. Sorrentino G, Ruggeri N, Specchia V, Cordenonsi M, Mano M, Dupont S, Manfrin A, Ingallina E, Sommaggio R, Piazza S, Rosato A, Piccolo S, Del Sal G. Metabolic control of YAP and TAZ by the mevalonate pathway. Nat Cell Biol. 2014;16(4):357–66.
84. Xin M, Kim Y, Sutherland LB, Murakami M, Qi X, McAnally J, Porrello ER, Mahmoud AI, Tan W, Shelton JM, Richardson JA, Sadek HA, Bassel-Duby R, Olson EN. Hippo pathway effector Yap promotes cardiac regeneration. Proc Natl Acad Sci U S A. 2013;110(34):13839–44.
85. Heallen T, Zhang M, Wang J, Bonilla-Claudio M, Klysik E, Johnson RL, Martin JF. Hippo pathway inhibits Wnt signaling to restrain cardiomyocyte proliferation and heart size. Science. 2011;332(6028):458–61.
86. Maejima Y, Kyoi S, Zhai P, Liu T, Li H, Ivessa A, Sciarretta S, Del Re DP, Zablocki DK, Hsu CP, Lim DS, Isobe M, Sadoshima J. Mst1 inhibits autophagy by promoting the interaction between Beclin1 and Bcl-2. Nat Med. 2013;19(11):1478–88.
87. Del Re DP, Matsuda T, Zhai P, Maejima Y, Jain MR, Liu T, Li H, Hsu CP, Sadoshima J. Mst1 promotes cardiac myocyte apoptosis through phosphorylation and inhibition of Bcl-xL. Mol Cell. 2014;54(4):639–50.

88. Wilkinson DS, Jariwala JS, Anderson E, Mitra K, Meisenhelder J, Chang JT, Ideker T, Hunter T, Nizet V, Dillin A, Hansen M. Phosphorylation of LC3 by the Hippo kinases STK3/STK4 is essential for autophagy. Mol Cell. 2015;57(1):55–68.
89. Singh A, Ramesh S, Cibi DM, Yun LS, Li J, Li L, Manderfield LJ, Olson EN, Epstein JA, Singh MK. Hippo signaling mediators yap and taz are required in the epicardium for coronary vasculature development. Cell Rep. 2016;15(7):1384–93.
90. Ramjee V, Li D, Manderfield LJ, Liu F, Engleka KA, Aghajanian H, Rodell CB, Lu W, Ho V, Wang T, Li L, Singh A, Cibi DM, Burdick JA, Singh MK, Jain R, Epstein JA. Epicardial YAP/TAZ orchestrate an immunosuppressive response following myocardial infarction. J Clin Invest. 2017;127(3):899–911.

# Chapter 8
# The Role of Autophagy in Mesenchymal Stem Cell-Based Suppression of Immune Response

**Vladislav Volarevic, Jelena Jakovljevic, C. Randall Harrell, Crissy Fellabaum, and Nebojsa Arsenijevic**

**Abstract** Mesenchymal stem cells (MSCs) are, due to their capacity for differentiation, immunomodulatory and proangiogenic characteristics, widely used as new therapeutic agents for the treatment of autoimmune, ischemic and degenerative diseases. One of the major barriers for successful transplantation of MSCs is their poor survival after engraftment in the inflamed and hypoxic tissues. Since autophagy regulates survival, differentiation potential, immunomodulatory and proangiogenic characteristics of engrafted MSCs, modulation of autophagy in transplanted MSCs may represent a novel strategy to improve MSCs-based therapy. Until now, modulation of autophagy as a new approach for enhancement of functional characteristics of MSCs has been examined in animal models of multiple sclerosis, osteoporosis, diabetes, myocardial infarction, and graft-versus-host disease. Obtained results suggest that regulation of autophagy may represent a new therapeutic approach that will enhance the efficacy of MSC-based therapy.

## Abbreviations

| | |
|---|---|
| 3-MA | 3-Methyladenine |
| aGVHD | Acute graft-versus-host disease |
| Ang-1 | Angiopoietin-1 |
| AT | Adipose tissue |
| ATP | Adenosine triphosphate |
| BM | Bone marrow |
| BMT | Bone marrow transplantation |
| CNS | Central nervous system |

V. Volarevic (✉) · J. Jakovljevic · N. Arsenijevic
Faculty of Medical Sciences, Department of Microbiology and immunology, Center for Molecular Medicine and Stem Cell Research, University of Kragujevac, Serbia, Kragujevac, Serbia

C. R. Harrell · C. Fellabaum
Regenerative Processing Plant, LLC, Palm Harbor, FL, USA

 
K. Turksen (ed.), *Autophagy in Health and Disease*, Stem Cell Biology and Regenerative Medicine, https://doi.org/10.1007/978-3-319-98146-8_8

| | |
|---|---|
| DCs | Dendritic cell |
| EAE | Experimental autoimmune encephalomyelitis |
| ECM | Extracellular matrix |
| ECs | Endothelial cells |
| EGF | Epidermal growth factor |
| EPCs | Endothelial progenitor cells |
| FGF-2 | Fibroblast growth factor |
| GIOP | Glucocorticoid-induced osteoporosis |
| HGF | Hepatic growth factor |
| HIF-1 | Hypoxia-inducible factor 1 |
| HLA | Human leukocyte antigen |
| HLA-G | Human leukocyte antigen-G |
| HO-1 | Heme oxygenase-1 |
| IDO | Indolamine 2,3-dioxygenase |
| IFN-γ | Interferon gamma |
| IL | Interleukin |
| IL-1Ra | IL-1 receptor antagonist |
| IL-6 | Interleukin-6 |
| LIF | Leukocyte inhibitory factor |
| LPS | Lipopolysaccharide |
| MCP-1 | Monocyte chemoattractant protein-1 |
| MHC | Major histocompatibility complex |
| MI | Myocardial infarction |
| miRNAs | microRNAs |
| MMPs | Matrix metalloproteinases |
| MS | Multiple sclerosis |
| MSCs | Mesenchymal stem cells |
| mTOR | Mammalian target of rapamycin |
| NK | Natural killer |
| NKT | Natural killer T |
| NO | Nitric oxide |
| PD-1 | Programmed death 1 |
| PGE2 | Prostaglandin E2 |
| PLGF | Placental growth factor |
| ROS | Reactive oxigene species |
| TGF-α | Transforming growth factor α |
| TGF-β | Transforming growth factor-beta |
| TGF-β | Transforming growth factor-β |
| TNF-α | Tumor necrosis factor alpha |
| TSG-6 | Tumor necrosis factor α-stimulated gene 6 |
| UCB | Umbilical cord blood |
| VEGF | Vascular endothelial growth factor |

## 8.1 Introduction

Autophagy, as an evolutionary conserved process, plays an important role in homeostasis and embryogenesis, functioning both as a survival and cell death pathway [1]. However, despite its importance in cell physiology, there is little information about the role of autophagy in stem cell survival and function. Although autophagy is sometimes associated with cell death, it is generally considered to be a survival mechanism because autophagy, unlike apoptosis and necrosis, is activated in conditions of stress represented by hypoxia, nutrient deprivation, and metabolic, oxidative, and proteotoxic stress [2]. Therefore, it has been argued that autophagy is a crucial cellular pathway that regulates development, differentiation, survival, and homeostasis of adult stem cells [3]. Accordingly, herewith, the effects of autophagy on stemness, survival, immunomodulatory and therapeutic characteristics of mesenchymal stem cells (MSCs), the most commonly used adult stem cells in clinical trials, has been emphasized.

## 8.2 Mesenchymal Stem Cells: New Players in Regenerative Medicine

MSCs (also known as multipotent mesenchymal stromal cells) were first described as fibroblast-like bone marrow populating cells by Friedenstein and coworkers [4]. Due to their immunomodulatory and proangiogenic ability, self-renewal, and differentiation capacity, MSCs are becoming new and promising therapeutic agents for the treatment of autoimmune, ischemic and degenerative diseases [5]. These cells can differentiate into all cell types of mesodermal origin and due to their plasticity, they are able to, in vitro, differentiate into cells of neuroectodermal (neurons, astrocytes, oligodendrocytes, epithelial cells) or endodermal (hepatocytes) origin [6].

Among stem cells, MSCs have several advantages for therapeutic use such as ability to migrate to the sites of tissue injury, strong immunosuppressive effects, better safety after infusion of allogeneic MSCs, and lack of ethical issues, such as those related to the application of human embryonic stem cells [7]. Simple acquisition, rapid proliferation, maintenance of differentiation potential after repeated passages in vitro, minor immunological rejection due to the low surface expression of major histocompatibility complex (MHC) antigens, efficient engraftment and long-term coexistence in the host are the main characteristics of MSCs that enable their therapeutic use [5–7]. Accordingly, the past decade has witnessed an extraordinary scientific production focused toward the possible clinical use of MSCs in the therapy of autoimmune, ischemic, and chronic inflammatory diseases with very promising findings [5–9].

As with other multipotent stem cells, MSCs have a high capacity for self-renewal while maintaining multipotency. The exact nature and localization of MSCs in vivo remain undefined, but it appears that they reside in almost all postnatal organs and tissues [8]. Apart from bone marrow (BM), MSCs or MSC-like cells have also been isolated from skeletal muscle, adipose tissue (AT), umbilical cord, blood (UCB) synovium, blood vessel walls, dental pulp, amniotic fluid as well as fetal blood, liver, and lungs [9]. BM, UCB, and AT have been most usually used as sources for the isolation of MSCs [10]. Differences between UCB-MSCs and other MSCs could be observed concerning the success rate of isolating, proliferation capacity and clonality. In contrast to BM-MSCs and AT-MSCs, UCB-MSCs have the highest rates of cell proliferation and clonality and significantly lower expression of p53, p21, and p16, well-known markers of senescence [10].

Diverse antigens have been found on the surface of MSC, but none of them appear to be unique for MSCs [7]. MSCs express CD105 (endoglin, also identified as SH2, a component of the receptor complex of transforming growth factor-beta (TGF-β) involved in proliferation, differentiation, and migration), CD73 (SH3/4, ectoenzyme that regulates the purinergic signaling through the hydrolysis of adenosine triphosphate (ATP)), CD44 (hyaluronan receptor involved in migration), CD90 (Thy-1, regulates differentiation of MSCs), stromal antigen 1 (involved in MSC migration), CD166 (vascular cell adhesion molecule), CD54/CD102 (intracellular adhesion molecule), and CD49b (Integrin alpha-2, involved in adhesion and osteogenic differentiation of MSCs) [10]. The International Society for Cellular Therapy has proposed minimal criteria to define human MSCs: (a) the cells must adhere to plastic in standard culture conditions using tissue culture flasks; (b) more than 95% of the cell population must express CD105, CD73, and CD90 but must lack expression (<2% positive) of CD45, CD34, CD14 or CD11b, CD79a or CD19, and HLA class II which are expressed on leukocytes, endothelial cells and trombocytes; and (c) the cells must be able to differentiate into adipocytes, osteoblasts, and chondrocytes under standard in vitro differentiating conditions [11].

MSC are a heterogenous population consisting of cells with variable growth potential, distinct morphologic and functional characteristics, but all of them have immunomodulatory and proangiogenic characteristics, representing a powerful tool in transplantational and regenerative medicine [12].

## 8.3 Molecular and Cellular Interactions Between MSCs and Immune Cells

The most intriguing aspect of the biology of MSCs is their immunomodulatory potential, such as capacity to suppress T cell proliferation and activation, dendritic cell (DCs) maturation and function, polarization of macrophages, suppression of B cell proliferation and differentiation in antibody producing plasma cells, and attenuation of cytotoxicity of natural killer (NK) and natural killer T (NKT) cells [13].

MSCs modulate proliferation, activation and function of immune cells in cell-to-cell contact (juxtacrine manner) or through the production of soluble factors (paracrine manner) [13]. Interraction between inhibitory molecule programmed death 1

(PD-1) with its ligands PD-L1 and PD-L2 is responsible for MSC-mediated suppression of T-cell proliferation [8, 13]. Despite this mechanism, the capacity of MSC to alter immune response is largely due to the production of soluble factors such as: transforming growth factor-β (TGF-β), hepatic growth factor (HGF), nitric oxide (NO), indolamine 2,3-dioxygenase (IDO), interleukin (IL)-10, IL-6, leukocyte inhibitory factor (LIF), IL-1 receptor antagonist (IL-1Ra), galectins, tumor necrosis factor α-stimulated gene 6 (TSG-6), human leukocyte antigen-G (HLA-G), heme oxygenase-1 (HO-1) and prostaglandin E2 (PGE2) [9, 13]. Through the production of TGF-β, MSCs suppress activation of Jak-Stat signaling pathway and cause the G1 cell cycle arrest, attenuating T, NK, and NKT cell proliferation [9]. IDO, an enzyme induced by proinflammatory cytokines (particularly interferon gamma (IFN-γ) and tumor necrosis factor alpha (TNF-α)) converts tryptophan to kynurenine. The degradation of tryptophan, amino acid that is essential for lymphocyte proliferation, has been suggested to inhibit lymphocyte proliferation [9]. Through the production of IDO, MSCs suppress proliferation of T lymphocytes and attenuates cytotoxicity and cytokine production in NKT cells [14–17]. Immediately after tissue injury, tissue resident macrophages produce inflammatory chemokines and cytokines which attract MSCs to the sites of wounding where they produce NO that suppress immune cells [16]. Additionally, MSC-derived NO can increase IDO activity and augment MSC-based suppression of immune response [17]. Through the secretion of PGE2, MSCs suppress IL-2 production and the expression of IL-2 receptor on T cells attenuating their proliferation and promote generation of immunosuppressive regulatory T cells [9, 13]. Additionally, MSC-derived PGE2 is responsible for inhibited maturation of DCs and increased alternative activation of macrophages [9].

MSCs modulate function of all immune cells affecting both innate and acquired immunity [13]. MSCs inhibit the division of stimulated T cells by preventing their entry into the S phase of the cell cycle and by mediating irreversible G0/G1 phase arrest [18]. In contrast to the strong inhibitory effects of MSCs on T cell proliferation, there are only relatively minor and reversible effects on T cell effector function, particularly IFN-γ production [19]. Additionally, MSCs do not significantly affect T cell activation (based on CD25 and CD69 surface expression on MSC-primed T cells), does not appear to be antigen specific, works across human leukocyte antigen (HLA) barriers and targets both primary and secondary T cell-driven immune response [19].

MSCs also inhibit the differentiation of monocytes into immature DCs blocking of the monocyte cell cycle at the G0 phase [20]. Cell contact between MSCs and DCs is not required for MSC-based modulation of DC maturation [13]. As a result of cross talk with DCs, MSCs produce soluble factors that attenuate maturation of DCs, downregulate expression of costimulatory molecules and suppress production of cytokines in DCs significantly reducing their ability to stimulate T cells [13].

B cell proliferation is inhibited by MSCs-derived soluble factors [9, 13]. B cell inhibition by MSCs is attributable to blockade of the G0/G1 phases of the cell cycle, similar to what occurs with T cells [21]. MSCs also reduce the expression of chemokine receptors and immunoglobulin production by activated B cells [21]. They do not, however, appear to alter surface molecules involved in stimulatory cell cooperation, such as HLA-DR, CD40 and the B7 costimulators, or to inhibit the production of TNF-α, IFN-γ, IL-4, and IL-10 [21].

MSCs significantly inhibit IL-2-stimulated proliferation of resting NK cells, but only partially impair proliferation of activated NK cells [13, 22]. Cell-to-cell contact and soluble factors such as TGF-β, NO, IDO, and PGE2 are responsible for this effect [9, 22]. Similarly, in NO and IDO dependent manner, MSCs suppress cytotoxicity and production of inflammatory cytokines in liver NKT cells [17, 23]. MSCs, through the secretion of immunosuppressive factors, suppress inflammatory M1 macrophages and promote their conversion in alternative M2, significantly attenuating macrophage driven inflammation [9, 13, 24].

## 8.4 Impact of MSCs on Angiogenesis

During ischemic injury, MSCs have been shown to promote angiogenesis through the: (a) secretion of trophic factors; (b) stimulation of endogenous endothelial progenitor cells (EPCs); (c) immune regulation of the microenvironment to enhance survival and proliferation of endothelial cells (ECs) [25].

A broad repertoire of angiogenic factors have been detected in the secretome of BM-MSCs, including vascular endothelial growth factor (VEGF), fibroblast growth factor (FGF)-2, angiopoietin-1 (Ang-1), monocyte chemoattractant protein-1 (MCP-1), interleukin-6 (IL-6), and placental growth factor (PLGF), that enhance angiogenesis [26–30]. Additionally, recently, in secretome of BM-MSCs, Cyr61 protein (cysteine-rich, angiogenic inducer 61), an important proangiogenic molecule, was identified [27] indicating its role in MSC-mediated generation of new blood vessels. It is well known that hypoxic conditions, TNF-α and lipopolysaccharide (LPS) upregulate the secretion of angiogenic factors by MSCs in an NF-kappa β dependent manner [29]. Transforming growth factor α (TGF-α), produced after ECs injury, through the activation of MEK/MAPK and the PI3K/AKT signaling pathways, induces production of proangiogenic factors (VEGF, HGF, IL-6, IL-8, and Ang-2) in BM-MSCs [30]. In order to grow, capillaries require degradation of the surrounding extracellular matrix (ECM) to allow endothelial cell sprouting. Matrix metalloproteinases (MMPs) are a group of enzymes that are responsible for the degradation of extracellular matrix proteins [29]. Some of them, including MMP2, MMP9, and MMP14 are secreted by MSCs [29], playing an important role in MSC-dependent modulation of angiogenesis.

Since MSCs can be found in the perivascular space in virtually all organs from where MSCs were obtained, particularly interesting is their interaction with pericytes and EPCs [31, 32]. Pericytes may be considered as vascular MSCs capable of migrating under appropriate stimulation from the MSC vascular niche to the vascular tube where they regulate the neovascularization by secreting cytokines, such as VEGF-A [33, 34]. EPCs stimulate angiogenesis mainly by secreting proangiogenic cytokines (VEGF, HGF, G-CSF, and IL-8) that induce recruitment, proliferation and survival of mature ECs [35]. MSC-EPC interaction relies on both paracrine and cell-to-cell communication [36]. Effects of MSCs and EPCs on angiogenesis are complementary since factors that are produced by EPCs successfully stimulate engraftment and MSC-mediated neovasculogenesis of transplanted MSCs [36]. Intercellular communication between MSCs pericytes and EPCs can also be modulated via the production of microcellular or nanocellular membrane vesicles, which can carry mediators (growth factors, cytokines, lipids, proteins) and genetic information (mRNA, premiRNA, miRNA, tRNA) between cells [37]. The ability of such microvesicles to stimulate angiogenesis has been described both in vitro and in vivo and emerging evidence indicates that microRNAs (miRNAs) play a significant role in MSC-mediated vascular biology and tissue repair [38].

## 8.5 The Role of Autophagy in the Maintenance of MSC Stemness, Survival, and Function

Appearing evidence indicates that autophagy plays a consistent role in the modulation of proliferation, differentiation and stemness of MSC [3]. Therefore, a great effort has been made trying to evaluate the role of autophagy induced by various extracellular or intracellular stimuli in the maintenance of MSC stemness, survival, and function [3].

In order to maintain their stemness, MSCs actively reduce a deterioration process by establishing low-reactive oxygen species environments [3]. Results from a recent study indicate that autophagy may have an important role in protecting stemness of MSCs from irradiation injury. It has been demonstrated that autophagy induced by starvation or rapamycin can reduce reactive oxygen species (ROS) accumulation-associated DNA damage, therefore maintaining stemness of MSC, whereas inhibition of autophagy leads to ROS accumulation and DNA damage, ultimately resulting in cell death [39].

Several studies investigated the role of hypoxia-induced autophagy for stemness, proliferation and survival of MSCs [40–43]. Hypoxia-inducible factor 1 (HIF-1) is a transcription factor, which functions as a master regulator of adaptive responses to hypoxia by improving a local microcirculation via its effects on vascular growth [40]. Recent studies have indicated that HIF-1 regulates the autophagy when MSCs are cultured under hypoxic conditions [41–43]. Hypoxia has been shown to promote BM-MSC proliferation, through the activation of apelin–APJ axis and through the activation of downstream autophagy pathway [41, 42]. Hypoxia-induced apoptosis of MSCs was increased by the autophagy inhibitor 3-methyladenine (3-MA), and decreased by rapamycin, a positive inducer of autophagy, suggesting that the self-eating process might protect MSCs from hypoxia-induced apoptosis. Additionally, atorvastatin, a commonly prescribed statin, could effectively activate autophagy via AMPK/mTOR pathway increasing survival of MSCs under hypoxic conditions [43].

Although results obtained in all these studies [40–43] strongly indicate beneficial effects of autophagy on stemness and survival of MSCs, opposite findings were recently reported [44, 45]. As demonstrated by Chang and colleagues, BM-MSCs cultured in a medium containing high glucose concentrations have premature senescence, genomic instability and telomere changes [44]. Activation of autophagy, correlated with senescence changes in BM-MSCs and deletion of autophagy-related genes accelerates senescence of MSCs [44, 45]. In line with this data, 3-MA-induced inhibition of autophagy prevents cell death of MSCs [44], suggesting that inhibition and not activation of autophagy was important for survival of MSCs that were cultured in conditions not optimal.

Commitment of transplanted and engrafted MSCs to different lineages is regulated by many cues in the local tissue microenvironment, such as plating density, cell shape, cytoskeleton tension and adhesive, mechanical or structural cellular properties, and is determined by a variety of growth factors [9]. Among these, FGF, epidermal growth factor (EGF) and HGF stimulate proliferation of engrafted MSC and promote their differentiation toward specific cell types [9]. In line with these findings, it has been recently shown that autophagy as well, can play an important role in the commitment of MSC to different lineages, especially toward the osteoblastic and adipogenic lineages [3]. It was demonstrated that undifferentiated MSC exist in a state of arrested autophagy with an accumulation of undegraded autophagic vacuoles and little autophagic turnover, whereas stimulation of osteogenic differentiation leads to a consistent increase in autophagis turnover [46]. Thus, autophagy seems to be of fundamental importance in the control of osteogenic differentiation and this seems to be related to the early mammalian target of rapamycin (mTOR) inhibition and the late activation of the Akt/mTOR signaling axis [3]. On the contrary, during adipogenic differentiation, the alteration in the autophagosome balance led to significant changes in differentiation efficiency of MSCs. Activation of autophagy inhibited adipocyte formation while accelerated fat accumulation was noticed after autophagosome blockade [46].

Autophagy also has an influence on the immunosuppressive function of MSCs. Most recently, Gao and coworkers demonstrated that rapamycin-induced activation of autophagy strengthened the capacity of MSCs to inhibit $CD4^+$ T cell proliferation, whereas 3-MA-induced inhibition of autophagy weakened the immunosuppressive potential of MSCs against effector T helper cells. It seems that autophagy mainly affected production of immunosuppressive TGF-β in MSCs modulating their capacity to inhibit CD4+ T cells. Rapamycin-pretreated MSCs secreted more while 3-MA-pretreated MSCs secreted less amounts of TGF-β when compared with the control MSCs. Also, exogenous addition of TGF-β recovered the immunosuppressive capacity of 3-MA-pretreated MSCs, whereas blocking of TGF-β (by anti-TGF-β monoclonal antibody) significantly reduced the immunosuppressive capacity of rapamycin-pretreated MSCs toward CD4+ T cells, indicating the importance of cross talk between TGF-β signaling and autophagy pathways for MSC-mediated inhibition of T cell-driven pathology [47]. However, completely opposite findings were reported by Dang and colleagues who investigated the role of autophagy in MSC-mediated suppression of CD4+ T cells in experimental autoimmune encephalomyelitis, murine model of multiple sclerosis [48]. They indicated that inhibition and not activation of autophagy is important for enhancement of MSC-mediated suppression of effector T cells. An inhibition of autophagy significantly increased MAPK 1/3 activation in MSCs, which was essential for PGE2-dependent suppression of CD4+ T cells and attenuation of EAE in MSC-treated animals [48].

In addition to its effects of immunomodulatory functions of MSCs, activation of autophagy in tumor-infiltrated MSCs may provide support for the growth of neighboring tumor cells [49]. MSCs, cultured in serum-free medium, survive prolonged serum deprivation by utilizing autophagy to recycle macromolecules in beclin-1, ATG10, ATG12 and MAP-LC3 dependent manner. At the same time, MSCs with upregulated autophagy-related genes, increased production of antiapoptotic factors (insulin-like growth factor 1, insulin-like growth factor 2, TGF-β, and insulin-like growth factor binding protein 2) that facilitated the survival of surrounding tumor cells enabling rapid tumor growth [49]. Among all these MSC-derived antiapoptotic factors, it seems that TGF-β had the most important role since TGF-β neutralization completely abrogated protective effects of autophagy activated MSC-derived secretome on tumor cell survival [49].

## 8.6 The Effects of Autophagy on the Therapeutic Potential of MSCs

One of the major barriers for successful transplantation of MSCs is their poor survival after engraftment in the inflamed and hypoxic tissues [3]. Since autophagy regulates survival [40–43], differentiation potential [46] and immunomodulatory characteristics [47, 48] of engrafted MSCs, modulation of autophagy in transplanted MSCs may represent a novel strategy to improve MSCs-based therapy of autoimmune, ischemic and degenerative diseases [3]. Until now, modulation of autophagy as new approach for enhancement of immunomodulatory properties of MSCs has been examined in animal models of multiple sclerosis (MS), osteoporosis, diabetes, myocardial infarction, and graft-versus-host disease.

MS is an autoimmune disease that is characterized by inflammation of central nervous system (CNS), demyelination, axonal loss, and degeneration. Results obtained in EAE, well established murine model of MS, indicated the crucial importance of myelin-specific T cells in the pathogenesis of this disease. These CD4+ T cells, activated on periphery, migrate and infiltrate into the CNS, where they, through the production of inflammatory cytokines (TNFα, IFN-γ, IL-17A) induce damage of the myelin and axons [50]. MSC-mediated suppression of CNS-infiltrated CD4+T cells has been shown as promising cell based therapy for the treatment of MS [50]. However, only paucity of transplanted MSCs have been noticed in CNS of MSC-treated mice suffering from EAE indicating that modulation of autophagy should be tested as new approach for enhancement of MSC survival in CNS. It is well known that MSCs, engrafted in CNS, enhance autophagy of neighboring neuronal cells having neuroprotective effect [51]. Additionally, in response to inflammatory cytokines (TNF-α and IFN-γ), MSCs that were transplanted in mice with EAE undergo autophagy, as well, by inducing expression of Beclin 1. Activation of autophagy significantly attenuates capacity of MSCs to suppress CD4+ T cells while inhibition of autophagy (by knocking down Beclin 1), significantly improved the therapeutic effects of MSCs on EAE increasing their immunosuppressive effects on $CD4^+$ T cells [48].

Glucocorticoid-induced osteoporosis (GIOP) is a widespread clinical complication of glucocorticoid therapy, and the most common type of secondary osteoporosis [52]. Oral glucocorticoids reduce the proliferation and increase the apoptosis of osteoblasts, prolong the survival of osteoclasts and enhance bone resorption, so the risk of bone fracture is increased in patients that receive glucocorticoid therapy. At the same time, BM-MSCs represent key cellular source for bone repair and regeneration in GIOP patients [52]. Recently, autophagy was determined as an important mechanism responsible for maintenance of bone tissue homeostasis in GIOP due to its effects on survival of BM-MSCs [52]. Glucocorticoid therapy induced autophagy in BM-MSCs as a mechanism of protection from starvation-induced apoptosis. Accordingly, 3-MA-induced inhibition of autophagy reduced proliferation of BM-MSCs and increased apoptosis of BM-MSCs resulting with the reduction in bone mass. These findings strongly suggest that regulation of autophagy should be considered as a new strategy aimed to increase effects of BM-MSC-based therapy in GIOP patients [52].

Due to their immunomodulatory and proangiogenic characteristics BM-MSCs have been widely used as an attractive candidate for cell-based therapy of myocardial infarction (MI) [53]. However, previous studies have revealed that BM-MSCs have undergone an acute death in 1 week after transplantation in the infarcted heart [53]. Since poor viability of engrafted MSCs limits their therapeutic efficiency, new approaches that will enhance viability of transplanted MSCs are urgently needed. A recently published study has shown that survival of MSCs after transplantation in damaged myocardium can be enhanced by drugs like atorvastatin, which activates autophagy via the AMPK/mTOR pathway [41]. Activation of autophagy in MSCs enables their survival under hypoxic conditions [41]. Additionally, apoptosis of BM-MSCs under hypoxic condition was regulated by autophagy and AMPK/mTOR pathway, as well, indicating that activation of autophagy may be useful approach to enhance survival of engrafted MSCs in ischemic myocardium [54].

The possible therapeutic effect of MSCs in diabetes is suggested by their capacity to generate insulin-producing cells in vitro and to normalize hyperglycemia in vivo, in a diabetic animal model [10]. Additionally, several lines of evidence suggested that mainly due to their proangiogenic characteristics, MSCs may be used for the treatment of diabetic complications: lower limb ischemia, polyneuropathy, cardiomyopathy, nephropathy, erectile dysfunction [5]. However, as previously discussed, MSCs transplanted into an ischemic environment have reduced cell survival rates and impaired angiogenic capacity. Therefore, pretreatment of MSCs in vitro has become a primary method to improve their survival and efficiency. Liu J and colleagues found that hypoxic pretreatment and elevated expression of HIF-1-α did not alter phenotype and differentiation potential of MSCs, but managed to significantly enhance their survival by promoting autophagy and by inhibiting apoptosis through the activation of AMPK/mTOR signaling pathway [55]. Furthermore, activated autophagy in MSCs correlated with increased angiogenesis in the lower limbs of MSC-treated ischemic diabetic rats [55], suggesting that induction of autophagy can be a useful approach for enhancement of proangiogenic characteristics of MSCs. Similar conclusions were made by Liu G and coworkers who evaluated induction of autophagy in MSCs as a new approach for the enhancement of their therapeutic potential in the therapy of diabetic erectile dysfunction [56]. Autophagy can be induced through the JNK-mediated phosphorylation or degradation of Bcl-2, which attenuates Bcl-2 dependent inhibition of Beclin-1 [57]. When MSCs engraft in inflammatory microenvironment, ROS induce, at the same time, apoptosis and autophagy in transplanted MSCs through JNK-mediated Bcl-2 degradation. An augmentation of autophagy counteracts apoptosis in MSCs, thus prolonging MSC survival and improving their therapeutic efficacy in the treatment of diabetic erectile dysfunction [56].

Acute graft-versus-host disease (aGVHD) remains a lethal and significant complication in allogeneic bone marrow transplantation (BMT) recipients [58]. MSCs can protect BMT recipients from the lethal aGVHD through the production of immunosuppressive factors (IL-10, TGF-β and IDO) [58]. However, it has to be highlighted that MSCs are not always immunosuppressive [13]. When MSCs are transplanted in the tissue with high levels of inflammatory cytokines, MSCs develop an immunosuppressive phenotype, but when MSCs are engrafted in the microenvironment with low levels of inflammatory mediators, they obtain proinflammatory phenotype, produce large amounts of proinflammatory cytokines and chemokines that stimulate activation and migration of neutrophils and T cells and increase inflammation [13]. These opposite actions may limit therapeutic use of MSCs in the treatment of aGVHD, and the optimization of their immunomodulatory properties can increase safeness of MSC-based therapy of aGVHD. Kim and colleagues demonstrated that rapamycin-induced activation of autophagy in AT-MSCs significantly increased expression of autophagy related genes (ATG5, LC3A and LC3B) that resulted with increased production of immunosuppressive factors (IL-10, IDO, and TGF-β). Additionally, by promoting expansion of T regulatory cells and by attenuating proliferation and effector functions of CD4+ Th17 cells, rapamycin-treated AT-MSCs more efficiently attenuated aGVHD than control AT-MSCs in vivo [58]. Therefore, activation of autophagy can be further explored as new approach for optimization of the immunomodulatory characteristics of AT-MSCs in the cell-based therapy of aGVHD patients.

## 8.7 Conclusions

Despite the fact that MSCs are, due to their differentiation, immunomodulatory, and proangiogenic properties, widely used for the treatment of autoimmune, ischemic, and degenerative diseases, these cells have undergone an acute death within 1 or few weeks after transplantation in the ischemic microenvironment. Since an augmentation of autophagy significantly enhances viability, immunosuppressive, and proangiogenic characteristics of engrafted MSCs, regulation of autophagy may represent a new therapeutic approach that will enhance the efficacy of MSC-based therapy.

**Acknowledgments** This work was supported by Serbian Ministry of Science (ON175069, ON175103) and Faculty of Medical Sciences University of Kragujevac (JP02/09).

## References

1. Mizushima N, Komatsu M. Autophagy: renovation of cells and tissues. Cell. 2011;147:728–41.
2. Denton D, Xu T, Kumar S. Autophagy as a pro-death pathway. Immunol Cell Biol. 2015;93:35–42.
3. Sbrana FV, Cortini M, Avnet S, Perut F, Columbaro M, De Milito A, Baldini N. the role of autophagy in the maintenance of stemness and differentiation of mesenchymal stem cells. Stem Cell Rev. 2016;12:621–33.
4. Friedenstein AJ, Chailakhjan RK, Lalykina KS. The development of fibroblast colonies in monolayer cultures of guinea-pig bone marrow and spleen cells. Cell Tissue Kinet. 1970;3:393–403.
5. Volarevic V, Arsenijevic N, Lukic ML, Stojkovic M. Concise review: mesenchymal stem cell treatment of the complications of diabetes mellitus. Stem Cells. 2011;29:5–10.
6. Markovic BS, Kanjevac T, Harrell CR, Gazdic M, Fellabaum C, Arsenijevic N, Volarevic V. Molecular and cellular mechanisms involved in mesenchymal stem cell-based therapy of inflammatory bowel diseases. Stem Cell Rev. 2018;14:153. https://doi.org/10.1007/s12015-017-9789-2.
7. Volarevic V, Ljujic B, Stojkovic P, Lukic A, Arsenijevic N, Stojkovic M. Human stem cell research and regenerative medicine--present and future. Br Med Bull. 2011;99:155–68.
8. Phinney DG, Prockop DJ. Concise review: mesenchymal stem/multipotent stromal cells: the state of transdifferentiation and modes of tissue repair—current views. Stem Cells. 2007;25:2896–902.
9. Volarevic V, Gazdic M, Simovic Markovic B, Jovicic N, Djonov V, Arsenijevic N. Mesenchymal stem cell-derived factors: immuno-modulatory effects and therapeutic potential. Biofactors. 2017;43:633–44.
10. Volarevic V, Al-Qahtani A, Arsenijevic N, Pajovic S, Lukic ML. Interleukin-1 receptor antagonist (IL-1Ra) and IL-1Ra producing mesenchymal stem cells as modulators of diabetogenesis. Autoimmunity. 2010;43:255–63.
11. Dominici M, Le Blanc K, Mueller I, Slaper-Cortenbach I, Marini F, Krause D, Deans R, Keating A, Prockop DJ, Horwitz E. Minimal criteria for defining multipotent mesenchymal stromal cells. The International Society for Cellular Therapy position statement. Cytotherapy. 2006;8:315–7.
12. Nurkovic J, Volarevic V, Lako M, Armstrong L, Arsenijevic N, Stojkovic M. Aging of stem and progenitor cells: mechanisms, impact on therapeutic potential, and rejuvenation. Rejuvenation Res. 2016;19:3–12.

13. Gazdic M, Volarevic V, Arsenijevic N, Stojkovic M. Mesenchymal stem cells: a friend or foe in immune-mediated diseases. Stem Cell Rev. 2015;11:280–7.
14. Milosavljevic N, Gazdic M, Simovic Markovic B, Arsenijevic A, Nurkovic J, Dolicanin Z, Jovicic N, Jeftic I, Djonov V, Arsenijevic N, Lukic ML, Volarevic V. Mesenchymal stem cells attenuate liver fibrosis by suppressing Th17 cells - an experimental study. Transpl Int. 2018;31:102. https://doi.org/10.1111/tri.13023.
15. Gazdic M, Arsenijevic A, Markovic BS, Volarevic A, Dimova I, Djonov V, Arsenijevic N, Stojkovic M, Volarevic V. Mesenchymal stem cell-dependent modulation of liver diseases. Int J Biol Sci. 2017;13:1109–17.
16. Simovic Markovic B, Gazdic M, Arsenijevic A, Jovicic N, Jeremic J, Djonov V, Arsenijevic N, Lukic ML, Volarevic V. Mesenchymal stem cells attenuate cisplatin-induced nephrotoxicity in iNOS-dependent manner. Stem Cells Int. 2017;2017:1315378.
17. Gazdic M, Simovic Markovic B, Vucicevic L, Nikolic T, Djonov V, Arsenijevic N, Trajkovic V, Lukic ML, Volarevic V. Mesenchymal stem cells protect from acute liver injury by attenuating hepatotoxicity of liver natural killer T cells in an inducible nitric oxide synthase- and indoleamine 2,3-dioxygenase-dependent manner. J Tissue Eng Regen Med. 2018;12:e1173. https://doi.org/10.1002/term.2452.
18. Glennie S, Soeiro I, Dyson PJ, Lam EW, Dazzi F. Bone marrow mesenchymal stem cells induce division arrest anergy of activated T cells. Blood. 2005;105:2821–7.
19. Yi T, Song SU. Immunomodulatory properties of mesenchymal stem cells and their therapeutic applications. Arch Pharm Res. 2012;35:213–21.
20. Ramasamy R, Fazekasova H, Lam EW, Soeiro I, Lombardi G, Dazzi F. Mesenchymal stem cells inhibit dendritic cell differentiation and function by preventing entry into the cell cycle. Transplantation. 2007;83:71–6.
21. Corcione A, Benvenuto F, Ferretti E, Giunti D, Cappiello V, Cazzanti F, Risso M, Gualandi F, Mancardi GL, Pistoia V, Uccelli A. Human mesenchymal stem cells modulate B-cell functions. Blood. 2006;107:367–72.
22. Gazdic M, Simovic Markovic B, Jovicic N, Misirkic-Marjanovic M, Djonov V, Jakovljevic V, Arsenijevic N, Lukic ML, Volarevic V. Mesenchymal stem cells promote metastasis of lung cancer cells by downregulating systemic antitumor immune response. Stem Cells Int. 2017;2017:6294717.
23. Milosavljevic N, Gazdic M, Simovic Markovic B, Arsenijevic A, Nurkovic J, Dolicanin Z, Djonov V, Lukic ML, Volarevic V. Mesenchymal stem cells attenuate acute liver injury by altering ratio between interleukin 17 producing and regulatory natural killer T cells. Liver Transpl. 2017;23:1040–50.
24. Simovic Markovic B, Nikolic A, Gazdic M, Nurkovic J, Djordjevic I, Arsenijevic N, Stojkovic M, Lukic ML, Volarevic V. Pharmacological inhibition of gal-3 in mesenchymal stem cells enhances their capacity to promote alternative activation of macrophages in dextran sulphate sodium-induced colitis. Stem Cells Int. 2016;2016:2640746.
25. Poggi A, Prevosto C, Massaro AM, Negrini S, Urbani S, Pierri I, Saccardi R, Gobbi M, Zocchi MR. Interaction between human NK cells and bone marrow stromal cells induces NK cell triggering: role of NKp30 and NKG2D receptors. J Immunol. 2005;175:6352–60.
26. Xia X, Tao Q, Ma Q, Chen H, Wang J, Yu H. Growth hormone-releasing hormone and its analogues: significance for MSCs-mediated angiogenesis. Stem Cells Int. 2016;2016:8737589.
27. Bronckaers A, Hilkens P, Martens W, Gervois P, Ratajczak J, Struys T, Lambrichts I. Mesenchymal stem/stromal cells as a pharmacological and therapeutic approach to accelerate angiogenesis. Pharmacol Ther. 2014;143:181–96.
28. Estrada R, Li N, Sarojini H, An J, Lee MJ, Wang E. Secretome from mesenchymal stem cells induces angiogenesis via Cyr61. J Cell Physiol. 2009;219:563–71.
29. Ghajar CM, Kachgal S, Kniazeva E, Mori H, Costes SV, George SC, Putnam AJ. Mesenchymal cells stimulate capillary morphogenesis via distinct proteolytic mechanisms. Exp Cell Res. 2010;316:813–25.

30. Crisostomo PR, Wang Y, Markel TA, Wang M, Lahm T, Meldrum DR. Human mesenchymal stem cells stimulated by TNF-alpha, LPS, or hypoxia produce growth factors by an NF kappa B- but not JNK-dependent mechanism. Am J Physiol Cell Physiol. 2008;294:C675–82.
31. Watt SM, Gullo F, van der Garde M, Markeson D, Camicia R, Khoo CP, Zwaginga JJ. The angiogenic properties of mesenchymal stem/stromal cells and their therapeutic potential. Br Med Bull. 2013;108:25–53.
32. Crisan M, Yap S, Casteilla L, Chen CW, Corselli M, Park TS, Andriolo G, Sun B, Zheng B, Zhang L, Norotte C, Teng PN, Traas J, Schugar R, Deasy BM, Badylak S, Buhring HJ, Giacobino JP, Lazzari L, Huard J, Péault B. A perivascular origin for mesenchymal stem cells in multiple human organs. Cell Stem Cell. 2008;3:301–13.
33. Caplan AI. All MSCs are pericytes? Cell Stem Cell. 2008;3:229–30.
34. Caplan AI. Why are MSCs therapeutic? New data: new insight. J Pathol. 2009;217:318–24.
35. de Souza LE, Malta TM, Kashima Haddad S, Covas DT. Mesenchymal stem cells and pericytes: to what extent are they related? Stem Cells Dev. 2016;25:1843–52.
36. Hur J, Yoon CH, Kim HS, Choi JH, Kang HJ, Hwang KK, Oh BH, Lee MM, Park YB. Characterization of two types of endothelial progenitor cells and their different contributions to neovasculogenesis. Arterioscler Thromb Vasc Biol. 2004;24:288–93.
37. De Luca A, Gallo M, Aldinucci D, Ribatti D, Lamura L, D'Alessio A, De Filippi R, Pinto A, Normanno N. Role of the EGFR ligand/receptor system in the secretion of angiogenic factors in mesenchymal stem cells. J Cell Physiol. 2011;226:2131–8.
38. Merino-González C, Zuñiga FA, Escudero C, Ormazabal V, Reyes C, Nova-Lamperti E, Salomón C, Aguayo C. Mesenchymal stem cell-derived extracellular vesicles promote angiogenesis: potencial clinical application. Front Physiol. 2016;7:24.
39. Hou J, Han ZP, Jing YY, Yang X, Zhang SS, Sun K, Hao C, Meng Y, Yu FH, Liu XQ, Shi YF, Wu MC, Zhang L, Wei LX. Autophagy prevents irradiation injury and maintains stemness through decreasing ROS generation in mesenchymal stem cells. Cell Death Dis. 2013;4:e844.
40. Taylor CT. Mitochondria and cellular oxygen sensing in the HIF pathway. Biochem J. 2008;409:19–26.
41. Li L, Li L, Zhang Z, Jiang Z. Hypoxia promotes bone marrow-derived mesenchymal stem cell proliferation through apelin/APJ/autophagy pathway. Acta Biochim Biophys Sin (Shanghai). 2015;47:362–7.
42. Molaei S, Roudkenar MH, Amiri F, Harati MD, Bahadori M, Jaleh F, Jalili MA, Mohammadi RA. Down-regulation of the autophagy gene, ATG7, protects bone marrow-derived mesenchymal stem cells from stressful conditions. Blood Res. 2015;50:80–6.
43. Zhang Q, Yang YJ, Wang H, Dong QT, Wang TJ, Qian HY, Xu H. Autophagy activation: a novel mechanism of atorvastatin to protect mesenchymal stem cells from hypoxia and serum deprivation via AMP-activated protein kinase/mammalian target of rapamycin pathway. Stem Cells Dev. 2012;21:1321–32.
44. Chang TC, Hsu MF, Wu KK. High glucose induces bone marrow-derived mesenchymal stem cell senescence by upregulating autophagy. PLoS One. 2015;10:e0126537.
45. Rubinsztein DC, Mariño G, Kroemer G. Autophagy and aging. Cell. 2011;146:682–95.
46. Nuschke A, Rodrigues M, Stolz DB, Chu CT, Griffith L, Wells A. Human mesenchymal stem cells/multipotent stromal cells consume accumulated autophagosomes early in differentiation. Stem Cell Res Ther. 2014;5:140.
47. Gao L, Cen S, Wang P, Xie Z, Liu Z, Deng W, Su H, Wu X, Wang S, Li J, Ouyang Y, Wu Y, Shen H. Autophagy improves the immunosuppression of CD4+ T cells by mesenchymal stem cells through transforming growth factor-β1. Stem Cells Transl Med. 2016;5:1496–505.
48. Dang S, Xu H, Xu C, Cai W, Li Q, Cheng Y, Jin M, Wang RX, Peng Y, Zhang Y, Wu C, He X, Wan B, Zhang Y. Autophagy regulates the therapeutic potential of mesenchymal stem cells in experimental autoimmune encephalomyelitis. Autophagy. 2014;10:1301–15.
49. Sanchez CG, Penfornis P, Oskowitz AZ, Boonjindasup AG, Cai DZ, Dhule SS, Rowan BG, Kelekar A, Krause DS, Pochampally RR. Activation of autophagy in mesenchymal stem cells provides tumor stromal support. Carcinogenesis. 2011;32:964–72.

50. Gharibi T, Ahmadi M, Seyfizadeh N, Jadidi-Niaragh F, Yousefi M. Immunomodulatory characteristics of mesenchymal stem cells and their role in the treatment of multiple sclerosis. Cell Immunol. 2015;293:113–21.
51. Shin JY, Park HJ, Kim HN, Oh SH, Bae JS, Ha HJ, Lee PH. Mesenchymal stem cells enhance autophagy and increase β-amyloid clearance in Alzheimer disease models. Autophagy. 2014;10:32–44.
52. Wang L, Fan J, Lin YS, Guo YS, Gao B, Shi QY, Wei BY, Chen L, Yang L, Liu J, Luo ZJ. Glucocorticoids induce autophagy in rat bone marrow mesenchymal stem cells. Mol Med Rep. 2015;11:2711–6.
53. Henning RJ. Stem cells in cardiac repair. Future Cardiol. 2011;7:99–117.
54. Zhang Z, Yang M, Wang Y, Wang L, Jin Z, Ding L, Zhang L, Zhang L, Jiang W, Gao G, Yang J, Lu B, Cao F, Hu T. Autophagy regulates the apoptosis of bone marrow-derived mesenchymal stem cells under hypoxic condition via AMP-activated protein kinase/mammalian target of rapamycin pathway. Cell Biol Int. 2016;40:671–85.
55. Liu J, Hao H, Huang H, Tong C, Ti D, Dong L, Chen D, Zhao Y, Liu H, Han W, Fu X. Hypoxia regulates the therapeutic potential of mesenchymal stem cells through enhanced autophagy. Int J Low Extrem Wounds. 2015;14:63–72.
56. Liu GY, Jiang XX, Zhu X, He WY, Kuang YL, Ren K, Lin Y, Gou X. ROS activates JNK-mediated autophagy to counteract apoptosis in mouse mesenchymal stem cells in vitro. Acta Pharmacol Sin. 2015;36:1473–9.
57. Wei Y, Pattingre S, Sinha S, Bassik M, Levine B. JNK1-mediated phosphorylation of Bcl-2 regulates starvation-induced autophagy. Mol Cell. 2008;30:678–88.
58. Kim KW, Moon SJ, Park MJ, Kim BM, Kim EK, Lee SH, Lee EJ, Chung BH, Yang CW, Cho ML. Optimization of adipose tissue-derived mesenchymal stem cells by rapamycin in a murine model of acute graft-versus-host disease. Stem Cell Res Ther. 2015;6:202.

# Chapter 9
# Emerging Connections: Synaptic Autophagy in Brain Aging and Disease

**YongTian Liang**

**Abstract** The imbalance of proteostasis has been implicated in brain aging and neurodegenerative diseases. Therefore, clearing dysfunctional proteins and organelles in neurons via macroautophagy opens a new avenue to rejuvenate the protein pools and, thus, promotes synaptic and neuronal integrity and function. Evidence shows that autophagy is crucial in regulating neuronal development and maintaining neuronal integrity. Recent work has demonstrated that autophagosome formation is prominent at synaptic terminals, stimulating research interest in the process of synaptic autophagy. Furthermore, the roles for autophagosomes in transfering neuronal signaling during their retrograde transport to the soma, maintaining neuronal homeostasis and synaptic plasticity are beginning to emerge, yet we are only at the inception of our understanding of synapse-specific regulatory factors involved in synaptic autophagy. Hence, delineating interactions between synaptic cargoes and synaptic autophagy will provide a more comprehensive understanding of the roles for autophagy in maintaining neuronal function by regulating synaptic transmission and plasticity. In this chapter, I will briefly review how synaptic autophagy intersects with brain aging and disease.

## Abbreviations

| | |
|---|---|
| AD | Alzheimer's disease |
| Alfy | Autophagy linked FYVE protein |
| ALS | Amyotrophic lateral sclerosis |
| Ambra1 | Activating molecule in Beclin1-regulated autophagy |
| AMI | Age-induced memory impairment |
| ARM | Anesthesia-resistant memory |
| ASD | Autism spectrum disorders |

Y. Liang (✉)
Institute of Biology/Genetics, Freie Universität Berlin, Berlin, Germany

NeuroCure, Cluster of Excellence, Charité Universitätmedizin, Berlin, Germany
e-mail: yongtian.liang@fu-berlin.de



K. Turksen (ed.), *Autophagy in Health and Disease*, Stem Cell Biology and Regenerative Medicine, https://doi.org/10.1007/978-3-319-98146-8_9

| | |
|---|---|
| ASM | Anesthesia-sensitive memory |
| Aß | Amyloid-ß |
| AZs | Active zones |
| BDNF | Brain-derived neurotrophic factor |
| CMA | Chaperone-mediated autophagy |
| CNS | Central nervous system |
| DR | Dietary restriction |
| eIF5A | Eukaryotic translation initiation factor 5A |
| FTLD | Frontotemporal lobar degeneration |
| HD | Huntington's disease |
| HTT | *Huntingtin* |
| MFRTA | Mitochondrial free radical theory of aging |
| PD | Parkinson's disease |
| PN | Proteostasis network |
| STED | stimulated emission depletion |
| SV | Synaptic vesicles |
| UPS | Ubiquitin–proteasome system |

## 9.1 Aging and Aging-Associated Diseases

Since time immemorial, kings and emperors have searched the longevity elixir for eternal life to reverse the aging process. Mankind's quest to live longer has been successful to a certain degree. With the advances in public health and medical conditions, life expectancy has been increasing considerably throughout the twentieth century. However, the potentially more important metric health span does not necessarily come along with extending life span. It is reassuring that healthy aging has been recognized at least as equally important as longer life span. Considerable evidence suggests that aging is one of the highest risk factors known for most human diseases and mortality [1]. Put differently, it is the increased threat for maladies such as cardiovascular disease, cancer, diabetes, sarcopenia, osteoporosis, and cognitive decline that is more troublesome than the endpoint [2]. Intuitively, it is now imperative that we discuss more intensively attributes of cellular aging and aging biochemistry as well as its impact on neurodegenerative diseases.

In fact, age-related disorders have already become a great socioeconomic burden around the world. Among them age-induced memory impairment (AMI) is deemed as one of the biggest challenges. In recent years, interventions and potential therapeutics that aim to improve life span and health span are being extensively investigated. Undeniably, the examinations of such novel interventions and/or therapeutics that act to slow the aging process will, in turn, help steer the field toward facilitating prolonged periods of healthiness and extending life span in humans.

Despite a growing body of literature and remarkable progress in aging research over the past few decades, the detailed mechanisms that drive aging and how the process of aging is regulated on a molecular and cellular level have yet to be fully elucidated. Astonishingly, over 300 theories of aging have been proposed to explain the events that lead to the death of an organism [3]. To complicate things further, they may interact with each other in a complex manner. Of note, the mitochondrial free radical theory of aging (MFRTA) remained arguably the main theory for aging for many decades. Yet this theory has been strongly challenged [4–6]. Thus far, modern theories of biological aging in humans mainly fall into two categories: programmed and random process, neither of which arrives at a full satisfaction. In brief, there is currently no consensus on these theories of aging, because the aging process is highly complex, involving multiple regulatory and executive mechanisms at different levels. Unsurprisingly, in the past few years it has been of great interest to understand which factors influence this inevitable and complex process. As a result, a wide array of molecular and cellular damages has been identified and shown to accumulate during aging. The lifelong accumulation of such damages will eventually result in frailty and diseases [7]. Thus, it remains a major challenge to identify and categorize factors that cause aging and their relative contribution to aging.

## 9.2 Aging and Neurodegenerative Diseases Converge on an Imbalance in Proteostasis Network

The cellular and molecular hallmarks that represent common denominators of aging have been proposed [8]. Aging is associated with a loss of proteostasis, altered nutrient sensing, organellar and mitochondrial damage, cellular senescence, and stem cell exhaustion, among other dysfunctions [8]. Lopez-Otin and collaborators identified the loss of protein homeostasis (proteostasis) and stem cell exhaustion as major processes involved in the decline of the regenerative potential capacity related to the buildup of age-associated damage [9]. Above all, loss of proteostasis was considered a primary hallmark, while stem cell decline emerged as an integrative hallmark [8].

Mechanistically, proteostasis is attained by the coordinated action of multiple cellular pathways known collectively as the proteostasis network (PN) [10], which regulates protein synthesis, folding, transport, quality control, and degradation to preserve the proteome integrity and limit the accumulation of protein aggregates [11]. Aging, disease-associated mutations, polymorphisms, and energetic deficits challenge the PN [12], and the accumulation of instable proteins gives rise to ubiquitous protein aggregation [13, 14] (see Fig. 9.1).

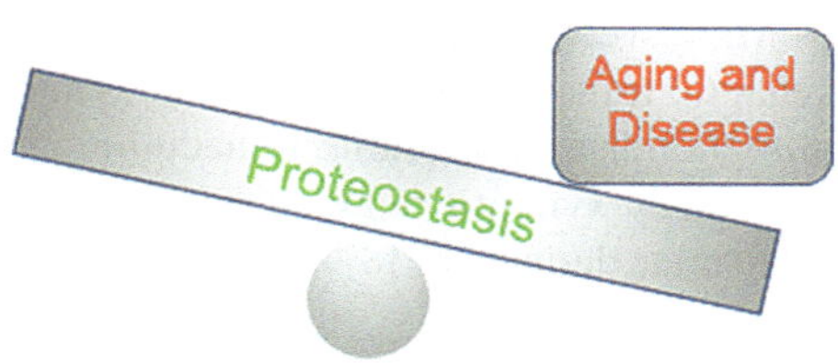


**Fig. 9.1** Aging and disease challenge the proteostasis network

Compelling evidence pinpoints that compromised or impaired proteostasis is major characteristics implicated in aging and neurodegenerative diseases [12, 15]. How might the impairment or even loss of proteostasis contribute to synaptic and neuronal dysfunction? Several studies have characterized the mouse brain synaptosomal proteome [16–18] and age-related proteomic changes in the brain [19–21].

Indeed, the accumulation of intracytoplasmic protein aggregations in neurons is a common hallmark implicated in many neurodegenerative diseases, including Parkinson's disease (PD), Huntington's disease (HD), amyotrophic lateral sclerosis (ALS), and Alzheimer's disease (AD) [22]. The etiology of PD remains elusive, but virtually all cases of PD involve the formation of intraneuronal presynaptic protein aggregates of α-synuclein [23], although the accrual of dysfunctional mitochondria and their turnover failure should not be neglected [24]. HD is caused by the *Huntingtin* (HTT) gene mutations which leads to polyglutamine (polyQ) expansion and, hence, neuronal proteostasis collapse [25, 26]. A common pathological hallmark of both ALS and frontotemporal lobar degeneration (FTLD) is the existence of cytoplasmic protein aggregates and inclusions in affected neurons and glia cells [27–29]. The most common dementia, namely AD, is characterized by the buildup of amyloid-ß (Aß) plaques and tau [22, 30–33].

In short, alterations in cellular proteostasis are a common feature of aging and many neurodegenerative disorders. The current challenge is to explore and identify the pathogenic mechanisms of neuronal proteostasis imbalance in different brain regions, and to understand how these proteostasis pathways ensure and confer neuronal and synaptic integrity.

## 9.3 Maintaining the Synaptic and Neuronal Integrity: A Formidable Challenge

Neurons are connected to each other via specialized contacts known as synapses, which are very dynamic structures that allow the relay of information between the presynaptic neuron and postsynaptic neuron via the release of neurotransmitters. These synaptic connections form neural networks, which is fundamental to diverse neural activities and brain functions, including motor action, sensory perception, learning, cognition, emotion, and sleep [34, 35]. Somewhat surprisingly, the fact that the "average" synapse may host around 300,000 proteins [36], which, beyond doubt, poses a great challenge for keeping the homeostasis of neuronal and synaptic proteome.

Moreover, increases of neuronal activities, especially energy-demanding synaptic transmission [37], render synaptic proteins and organelles (particularly mitochondria) vulnerable to accumulated damages. The intense synaptic activity places synaptic and neuronal proteome at high risk for misfolding and molecular damages [38].

Structurally, neurons are asymmetric cells that consist of somatodendritic and axonal compartments. This unique neuronal morphology complicates the long journey for protein transport from the soma to the presynaptic and postsynaptic specializations [39].

Studies in the adult human hippocampus show that neurogenesis occurs only very infrequently during adulthood, and then only within certain areas [40, 41]. Most synapses and neurons have to maintain their plasticity and integrity for the lifetime of an organism. Being post-mitotic, neurons simply cannot dilute defective proteins and organelles through cell division.

Collectively, these hostile circumstances raise the chance of an accumulation of defective proteins and/or organelles for extended periods of time. Intuitively, maintaining synaptic and neuronal integrity, while concurrently allowing for fine-tuning of synaptic plasticity constitute a formidable challenge [39]. Thus, one might expect specific mechanistic adaptations and solutions that support neurons in removing toxic proteins and defective organelles? Along these lines, it would be central to pinpoint how the mechanisms of synaptic protein homeostasis are tailored to the specific proteomic demands in neurons.

## 9.4 Autophagy Efficacy Critical for Neuronal Development and Integrity

In the course of evolution, neurons have utilized a number of adaptations and strategies to remove defective proteins in order to preserve the synaptic and neuronal homeostasis [12, 42]. Quality control mechanisms of the stability and functionality of the proteome include molecular chaperones and co-chaperones, the ubiquitin–proteasome system (UPS), and the autophagy machinery [12]. Notably, distinct forms of autophagy exist in the cell: macroautophagy (hereafter referred as autophagy), microautophagy, and chaperone-mediated autophagy (CMA) [38, 43–45]. Autophagy will be specifically focused in subsequent sections.

Autophagy is an evolutionarily conserved intracellular catabolic process that sequesters cytosolic proteins, membrane proteins, organelles, and protein aggregates into autophagosomes, which then fuse with lysosomes for bulk degradation. Selective autophagy recognizes and recruits targets via autophagy receptors. These targets have been described, such as aggregate-prone proteins (aggrephagy), damaged mitochondria (mitophagy), excess peroxisomes (pexophagy), ribosomes (ribophagy), endoplasmic reticulum (reticulophagy), and invading pathogens (xenophagy) [46].

The fundamental importance of autophagy is demonstrated by the fact that its dysregulation is implicated in multiple human diseases, including neurodegeneration, infection, and cancer [47, 48]. Multiple major diseases (including but not limited to age-associated and neurodegenerative pathologies) are being scrutinized for pathogenic aberrations in autophagy and their pharmacologic correction [49].

Clear-cut evidence pointed to the importance of autophagy in the developing central nervous system (CNS). Early studies established that a deficiency of *Atg7* or *Atg5* specifically in the brain causes neurodegeneration and the accumulation of polyubiquinated proteins as inclusion bodies [50, 51]. Fimia and colleagues found that Ambra1 (activating molecule in Beclin1-regulated autophagy) functional deficiency in mouse embryo results in severe neural tube defects related to impaired autophagy [52]. Dragich and colleagues reported that Alfy (autophagy linked FYVE protein) is required for axonal tracts formation throughout the brain and spinal cord [53]. Tang and colleagues found that developmental spine pruning is disrupted, whose deficits correlate with hyperactivation of mTOR (inhibiting autophagy). Moreover, neuronal specific autophagy-deficient mice exhibit autism spectrum disorders (ASD)-like behaviors and synaptic pathology [54]. Shedding more light on this issue, recent work has further demonstrated that restoration of autophagy in the brain suffices to rescue the neonatal lethality in *Atg5*-null mice [55].

It has been demonstrated that alterations in autophagy are also associated with neurodegeneration in humans [56–60]. Similarly, studies in different model organisms have implicated autophagy as a crucial regulator of aging [49, 61–63]. The last few years have witnessed an accural of evidence that efficacy of neuronal autophagy declines with age [62, 64–66]. In the aging human brain, autophagic core machinery, including ATG5 and ATG7, are transcriptionally downregulated with age [67].

Compromise in autophagic flux and efficacy seems causally involved in the progression of brain aging and neurodegeneration, but that neurons may upregulate autophagy to adapt and stimulate protective changes upon stress could be equally possible [39]. Interestingly, preliminary findings support the possibility that intercellular coordination of autophagy [68–70] may exist between neurons and glia in the brain. In addition, novel pathways may shuttle cellular garbage [71, 72] and even transfer signaling information from neurons to glia. Thus, the quality control pathways between neurons and glia might provide alternative mechanisms to support neuronal survival and function upon neurotoxic stress [73]. While the exact mechanistic adaptations of autophagy to the specific neuronal status must be further deciphered, autophagy, without doubt, operates as a critical checkpoint to assure the continuous renewal of intracellular proteins and organelles in neurons. Thus, the question arises whether promoting neuronal autophagy could protect from age-associated impairments?

## 9.5 Autophagy in Brain and Synapse Aging

A decade ago, elevating autophagy specifically in neurons was shown to extend life span in *drosophila* [74]. On the contrary, decreasing autophagy reduces life span and provokes neurodegeneration [74, 75]. Therefore, the restitution or promotion of autophagic function via different avenues has been proposed as one promising approach to delay aging, including brain aging [76].

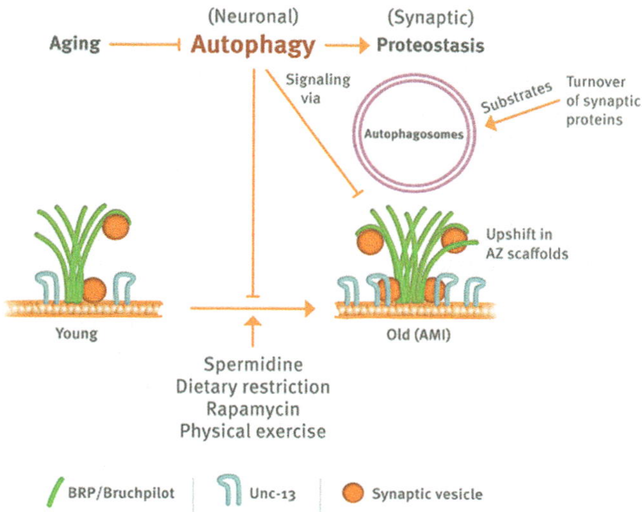


**Fig. 9.2** Neuronal autophagy and proteostasis intersecting with synaptic aging and cognitive decline. An age-associated upshift in the active zone scaffolds (Bruchpilot, Unc-13) size and release function might contribute to age-induced memory impairment (AMI). Antiaging treatments (spermidine and rapamycin supplement, dietary restriction, physical exercise) may rejuvenate autophagy and thus restore proteostasis. The retrograde transport of autophagosomes might also play a role in protective neuronal signaling (Modified from https://www.tandfonline.com/doi/full/10.1080/15548627.2016.1265193)

In this regard, treatments such as physical exercise, dietary restriction (DR), rapamycin, and spermidine have been tested in different context (see Fig. 9.2). Luo and colleagues reported that physical exercise preserves cognitive function via activated autophagy/mitophagy in the aged hippocampus, and lysosomal degradation is required for this process [77]. However, implementing healthy lifestyles, such as dietary restriction and/or physical exercise, might not be universally applicable. Thus, substances that may work as DR mimetics without (serious) side effects have spurred extraordinary interest. Rapamycin is found to extend life span across many species [78, 79] (see Fig. 9.2). Moreover, independent of its autophagy activation mechanism, rapamycin has been found to inhibit protein aggregates, an effect correlated with reduced protein translation [80, 81]. Chronic rapamycin treatment has been shown to improve cognitive performance [82]. However, rapamycin-based therapy suppresses immune function and might cause serious side effects when applied alone. Combination therapy with rapamycin or rapalogs might improve drug efficacy and delay the emergence of drug resistance, but further experiments exploring novel drug combinations with optimal doses should be carried out [83].

Is there body-endogenous substances that could reset autophagy at juvenile levels in aging tissues? Notably, polyamines were shown to decrease in an age-related fashion in a broad spectrum of animals and tissues investigated, and resubstitution of juvenile polyamine levels was shown to restore autophagy and thereby promote longevity in yeast, *C. elegans*, *Drosophila* and mice [84, 85]. While the exact mechanistic route linking spermidine levels to autophagy needs further attention, modulation of the proteome acetylation status via the inhibition of specific acetyltransferase might be involved [86]. Lately, regulation of autophagy at the level of translation has been revealed, where Lubas and colleagues discovered eukaryotic translation initiation factor 5A (eIF5A) hypusination promotes autophagy by enhancing ATG3 translation [87]. It is worth mentioning that eIF5A is the only known protein to contain spermidine-derived (spermidine transfers its aminobutyl moiety to one specific lysine residue of eIF5A precursor) hypusine [88]. Thus, it seems possible that spermidine-induced autophagy might, to some degree, stem from the hypusination of eIF5A.

*Drosophila* (the fruit flies) are a well-established powerful model used to study the mechanisms of aging and age-induced memory impairment [89–91], at least partially due to their short life span and easily manipulated genetics [90]. Intermediate-term memory can be further divided into anesthesia-sensitive memory (ASM) and anesthesia-resistant memory (ARM), with the latter being more stable [92]. The underlying mechanisms of AMI remain elusive, because when examining causative changes in brain structure it is hard to distinguish between adaptive and protective changes. Spermidine feeding was found to block AMI, namely age-sensitive memory component (ASM) which normally declines with age [93], while the age-insensitive component (ARM) was not altered (by spermidine). The protective effects in cognitive function are autophagy-dependent, as the loss of a single copy *Atg7* eliminated spermidine-mediated memory rescue in aged flies [62].

Studies in humans and animal model imply that impairments in cognitive function, including AMI, are not due to the loss of neurons, but linked to subtle age-associated alterations in synaptic plasticity [94]. Thus, a deeper understanding of AMI, especially at the synapse would ultimately deliver a conceptual framework for delaying and/or preventing AMI. Indeed, changes in the ultrastructural levels and, hence, "synapse strength" has been considered a key mechanism for memory formation [94, 95]. In our context, a brain-wide, age-dependent up-shift (termed RAMP-UP) was found in the ultrastructural size (in this context Bruchpilot and Unc-13) (see Fig. 9.2) and release function of the presynaptic active zones (AZs) in the fly brain from both electron microscopy and stimulated emission depletion (STED) microscopy quantifications [96] (see Fig. 9.2). This upshift may push synapses beyond their upper operational limit, narrow the "synaptic plasticity" process, and consequently interfere with learning and memory [97] (see Fig. 9.2).

In the aforementioned scenario, the synaptic basis for AMI, and the role of boosting autophagy via spermidine at the synapse in aged fruit flies exemplify how synaptic autophagy might tangle with cognitive function and many other functional aspects of the brain. These are only just emerging which should be subject to further scientific scrutiny. Obviously, synaptic autophagy has the potential to become of vital importance for protective strategies in the context of aging-associated dysfunctions of the brain. This brings up the question of what cargoes are degraded in neurons.

## 9.6 Emerging Synaptic Cargoes: Substrates for Synaptic Autophagy

So far, it remains ambiguous whether the presynaptic specializations (including synaptic proteins and synaptic vesicles), postsynaptic receptors, and synaptic organelles might be direct and/or specific substrates of the autophagic clearance program.

Several lines of evidence identify the molecular machinery (see Table 9.1) that controls autophagy at presynaptic terminals. Synaptic proteins such as EndophilinA [98] and Synaptojanin-1 [99] were reported to intersect with autophagy machinery to promote synaptic autophagy. Contrarily, a presynaptic scaffolding protein Bassoon [100] inhibits presynaptic autophagy by sequestering Atg5. These seemingly opposing regulation and coordination of presynaptic autophagy by synaptic proteins might serve as a dynamic switch during aging and neurodegenerative diseases [39].

Moreover, autophagosomal and endolysosomal mechanisms have been found to mediate synaptic vesicles (SV) degradation (see Table 9.1). Recently, it was shown that Rab26 links synaptic vesicles to the autophagy pathway [101]. Besides, synaptic endosomes act as sorting stations for synaptic vesicle proteins degradation, such as Rab5 [102] and Rab35 [103]. It is worth mentioning that autophagy and endolysosomal pathways are connected, as autophagosomes fuse with late endosomes prior to lysosomal degradation. Whether these two pathways work in parallel, or whether autophagy operates only as a complementary mechanism for SV degradation at the synapse remains unanswered.

Intriguingly, postsynaptic cargoes, including neurotransmitter receptors (see Table 9.1), can also be substrates for autophagic degradation. In 2006, Rowland and colleagues showed that GABAA receptors, but not acetylcholine receptors, can traffic to autophagosomes from the plasma membrane following removal by endocytosis [104]. Subsequently, Shehata and colleagues found that neuronal stimulation induces NMDA-receptor-dependent autophagy, which contributes to AMPA receptor degradation [105]. Furthermore, Nikoletopoulou and colleagues [106] reported that postsynaptic scaffold proteins, namely PICK1, PSD-95, and SHANK3 constitute substrates of autophagy. These studies, once again, hint at the possible role of autophagy in regulating synaptic plasticity.

Notably, selective degradation of damaged or unneeded mitochondria (see Table 9.1) by autophagy (mitophagy) might be crucial in maintaining cellular energetics under both basal and stressed situations at the synapse (see Fig. 9.3). Mitophagy

**Table 9.1** Emerging synaptic cargoes of autophagy

| Synaptic cargoes | Location | Synaptic and neuronal maintenance |
|---|---|---|
| Synaptic proteins | Pre- and postsynaptic | Regulate autophagy |
| Synaptic vesicles | Presynaptic | Regulate neurotransmission |
| Postsynaptic receptors | Postsynaptic | Regulate synaptic plasticity |
| Synaptic organelles | Presynaptic (e.g., mitochondria) | Maintain neuronal homeostasis |

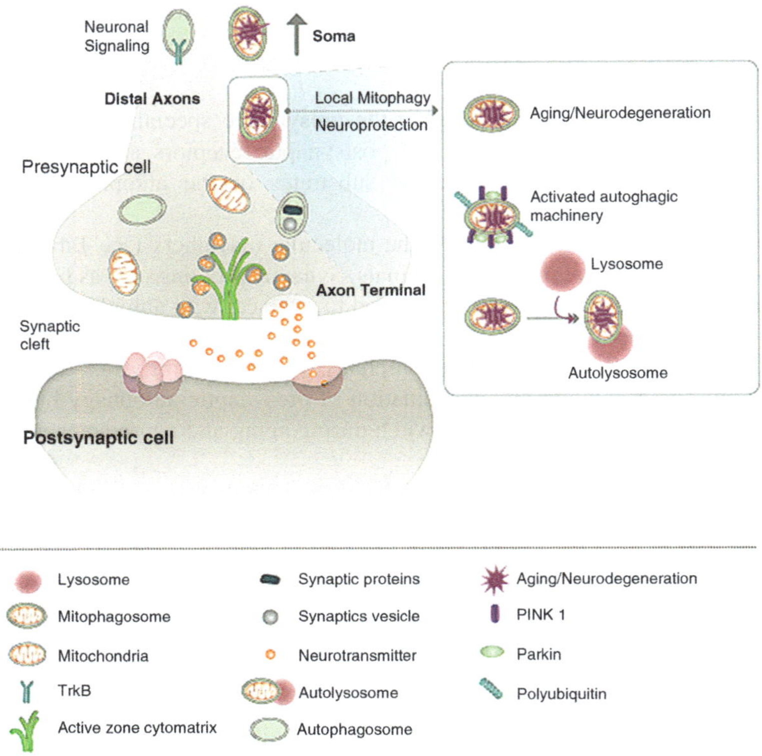


**Fig. 9.3** Neuronal signaling via the autophagosome retrograde transport to the soma and conferred neuroprotection through local mitophagy. (Source from https://www.sciencedirect.com/science/article/pii/S0959438817301241) Autophagosome formation occurs mainly in axons, and autophagosomes can mediate neuronal signaling during their retrograde transport to the soma. Autophagosomes are shown to engulf dysfunctional mitochondria and return them to the soma. Mitophagosomes fuse locally with lysosomes. Axonal mitophagy (zoom-in) likely provides quick neuroprotection against neurotoxic stress given the extreme neuronal structure and high local demand for ATP production posing formidable challenges for maintaining well-functional mitochondria

research has garnered much attention in the last decade, and some insights have been deciphered. The Youle group made a groundbreaking discovery that Parkin [107] and PINK1 [108] are recruited selectively to damaged or uncoupled mitochondria and promote their autophagy, providing evidence that links mitophagy and Parkinson's disease (see Fig. 9.3) [107, 108]. Cai [109] and Sung [110] found that Parkin-targeted mitochondria predominantly accumulate in the somatodendritic regions in neurons, which argued a gating mechanism might exist to allow only healthy mitochondria to be transported anterogradely. Perhaps not so surprisingly, Ashrafi [111] found that damaged mitochondria can be "eaten" locally within the mid-axon of primary hippocampal neurons and such local neuronal mitophagy is PINK1/Parkin-dependent. Overall, this finding supports the notion that mitophagy can take place in neuronal axons, and that local mitophagy could provide prompt neuroprotection by removing severely damaged mitochondria (see Fig. 9.3). However, it remains obscure how the homeostasis of presynaptic mitochondria is maintained in both physiological and pathological conditions, especially in the context of brain aging. Cao and colleagues [112] showed that axonal mitochondria are maintained independently of mitophagy, and that mitophagy-independent mechanisms such as fission–fusion may be central to maintain axonal mitochondrial integrity during normal aging. Indeed, the mechanisms of mitochondrial quality control are multitiered, operating at protein, suborganelle, organelle, and cell levels [113].

## 9.7 A Role for Autophagosomes in Neuronal Signaling and Synaptic Plasticity

By now, there is persuasive evidence that autophagosome biogenesis is very pronounced at presynaptic terminals. Neuronal autophagosome formation and maturation takes place mainly at the axonal compartment [114, 115].

Work from the Holzbaur lab and other labs showed that neuronal autophagosome biogenesis and maturation originate mainly in the axonal compartment [114–116]. Their follow-up study demonstrated that neurons show robust, constitutive autophagy with a marked augmentation of autophagosome formation in distal axons, though a minority of autophagosome can still be formed within the somatodendritic regions [117]. Overall, autophagy appears highly compartmentalized in primary neurons. Concerning the fact that different populations of autophagosomes may exist in different neuronal compartments, it is tempting to ask whether synaptic autophagy may execute distinct functions as to autophagy in the somatodendritic domain.

In contrast to other cell types and tissues where autophagy is induced upon starvation [118], autophagy shows no induction in the brain when following mTOR inhibition [117, 119, 120]. These outcomes indicate neuronal autophagy may primarily function to regulate neuronal development and maintain neuronal homeostasis [52–54, 121]. These surprising connections were manifested in recent work in the Haucke lab, where they demonstrated while neuronal autophagosomes undergo retrograde transport from the axon to the soma, they concurrently transport brain-derived neurotrophic factor (BDNF)-activated TrkB receptors (see Fig. 9.3), promoting neuronal branching and preventing neurodegeneration [122, 123]. Indeed, regulation of autophagy by fasting differs in distinct brain regions [106]. Recently, BDNF has been shown to suppress autophagy in cortical and hippocampal neurons, respectively [106, 124]. Moreover, the suppression of autophagy by BDNF might be required for synaptic plasticity [106].

Taken together, these findings reveal noncanonical functions of autophagosomes in neuronal signaling and synaptic plasticity, which are distinct from their canonical role in degradative pathways. Deciphering the mechanistic regulations of such "signaling autophagosomes" in details and various contexts is a forthcoming challenge and might open up a new therapeutic avenue for aging and diseased brain. One fascinating question is whether autophagosomes at the presynaptic and postsynaptic specializations have unique characteristics and/or regulatory mechanisms. Despite lingering questions, future studies are needed to disentangle and clarify the many roles autophagy plays on both sides of the synaptic cleft. A comprehensive understanding of the mechanisms and coordination of autophagy, other proteostasis pathways and homeostatic regulations at the synapse may provide new avenues for therapeutic intervention to delay brain aging and mitigate neurodegenerative diseases.

**Acknowledgments** I am deeply indebted to Stephan Sigrist for helpful discussion and resources. This work was supported by SFB958 of the Deutsche Forschungsgemeinschaft (DFG) and the China Scholarship Council (CSC).

## References

1. Armanios M, de Cabo R, Mannick J, Partridge L, van Deursen J, Villeda S. Translational strategies in aging and age-related disease. Nat Med. 2015;21(12):1395–9.
2. Granger A, Mott R, Emambokus N. Is aging as inevitable as death and taxes? Cell Metab. 2016;23(6):947–8.
3. Medvedecv AZ. An attempt at a rational classification of theories of ageing. Biol Rev. 1990;65(3):375–98.
4. Stuart JA, Maddalena LA, Merilovich M, Robb EL. A midlife crisis for the mitochondrial free radical theory of aging. Longevity Healthspan. 2014;3(1):4.
5. Lapointe J, Hekimi S. When a theory of aging ages badly. Cell Mol Life Sci. 2010;67(1):1–8.
6. Barja G. Updating the mitochondrial free radical theory of aging: an integrated view, key aspects, and confounding concepts. Antioxid Redox Signal. 2013;19(12):1420–45.
7. Kirkwood TB. A systematic look at an old problem. Nature. 2010;451(7179):644–7.
8. Lopez-Otin C, Blasco MA, Partridge L, Serrano M, Kroemer G. The hallmarks of aging. Cell. 2013;153(6):1194–217.
9. Revuelta M, Matheu A. Autophagy in stem cell aging. Aging Cell. 2017;16(5):912–5.
10. Powers ET, Morimoto RI, Dillin A, Kelly JW, Balch WE. Biological and chemical approaches to diseases of proteostasis deficiency. Annu Rev Biochem. 2009;78:959–91.
11. Sala AJ, Bott LC, Morimoto RI. Shaping proteostasis at the cellular, tissue, and organismal level. J Cell Biol. 2017;216(5):1231–41.
12. Labbadia J, Morimoto RI. The biology of proteostasis in aging and disease. Annu Rev Biochem. 2015;84:435–64.
13. David DC, Ollikainen N, Trinidad JC, Cary MP, Burlingame AL, Kenyon C. Widespread protein aggregation as an inherent part of aging in *C. elegans*. PLoS Biol. 2010;8(8):e1000450.
14. Demontis F, Perrimon N. FOXO/4E-BP signaling in *Drosophila* muscles regulates organism-wide proteostasis during aging. Cell. 2010;143(5):813–25.
15. Lim J, Yue Z. Neuronal aggregates: formation, clearance, and spreading. Dev Cell. 2015;32(4):491–501.
16. Schrimpf SP, Meskenaite V, Brunner E, Rutishauser D, Walther P, Eng J, Aebersold R, Sonderegger P. Proteomic analysis of synaptosomes using isotope-coded affinity tags and mass spectrometry. Proteomics. 2005;5(10):2531–41.
17. FM D, Birgit B, Stefan R, Larysa T, Giuseppina M, Turck CW. Profiling of mouse synaptosome proteome and phosphoproteome by IEF. Electrophoresis. 2010;31(8):1294–301.

18. Flynn JM, Czerwieniec GA, Choi SW, Day NU, Gibson BW, Hubbard A, Melov S. Proteogenomics of synaptosomal mitochondrial oxidative stress. Free Radic Biol Med. 2012;53(5):1048–60.
19. Poon HF, Vaishnav RA, Getchell TV, Getchell ML, Butterfield DA. Quantitative proteomics analysis of differential protein expression and oxidative modification of specific proteins in the brains of old mice. Neurobiol Aging. 2006;27(7):1010–9.
20. Mehan ND, Strauss KI. Combined age- and trauma-related proteomic changes in rat neocortex: a basis for brain vulnerability. Neurobiol Aging. 2012;33(9):1857–73.
21. Stauch KL, Purnell PR, Fox HS. Aging synaptic mitochondria exhibit dynamic proteomic changes while maintaining bioenergetic function. Aging. 2014;6(4):320–34.
22. Wang C, Telpoukhovskaia MA, Bahr BA, Chen X, Gan L. Endo-lysosomal dysfunction: a converging mechanism in neurodegenerative diseases. Curr Opin Neurobiol. 2018;48:52–8.
23. Dauer W, Przedborski S. Parkinson's disease: mechanisms and models. Neuron. 2003;39:889–909.
24. Gao F, Yang J, Wang D, Li C, Fu Y, Wang H, He W, Zhang J. Mitophagy in Parkinson's disease: pathogenic and therapeutic implications. Front Neurol. 2017;8:527.
25. Koyuncu S, Fatima A, Gutierrez-Garcia R, Vilchez D. Proteostasis of huntingtin in health and disease. Int J Mol Sci. 2017;18(7):pii: E1568.
26. Harding RJ, Tong Y-F. Proteostasis in Huntington's disease: disease mechanisms and therapeutic opportunities. Acta Pharmacol Sin. 2018;39:754.
27. Arai T, Hasegawa M, Akiyama H, Ikeda K, Nonaka T, Mori H, Mann D, Tsuchiya K, Yoshida M, Hashizume Y, Oda T. TDP-43 is a component of ubiquitin-positive tau-negative inclusions in frontotemporal lobar degeneration and amyotrophic lateral sclerosis. Biochem Biophys Res Commun. 2006;351(3):602–11.
28. Blokhuis AM, Groen EJ, Koppers M, van den Berg LH, Pasterkamp RJ. Protein aggregation in amyotrophic lateral sclerosis. Acta Neuropathol. 2013;125(6):777–94.
29. Gotzl JK, Lang CM, Haass C, Capell A. Impaired protein degradation in FTLD and related disorders. Ageing Res Rev. 2016;32:122–39.
30. Nijholt DA, de Graaf TR, van Haastert ES, Oliveira AO, Berkers CR, Zwart R, Ovaa H, Baas F, Hoozemans JJ, Scheper W. Endoplasmic reticulum stress activates autophagy but not the proteasome in neuronal cells: implications for Alzheimer's disease. Cell Death Differ. 2011;18(6):1071–81.
31. Morawe T, Hiebel C, Kern A, Behl C. Protein homeostasis, aging and Alzheimer's disease. Mol Neurobiol. 2012;46(1):41–54.
32. Villemagne VL, Okamura N. Tau imaging in the study of ageing, Alzheimer's disease, and other neurodegenerative conditions. Curr Opin Neurobiol. 2016;36:43–51.
33. Klaips CL, Jayaraj GG, Hartl FU. Pathways of cellular proteostasis in aging and disease. J Cell Biol. 2018;217(1):51–63.
34. Hering H, Sheng M. Dentritic spines: structure, dynamics and regulation. Nat Rev Neurosci. 2001;2:880.
35. Nimchinsky EA, Sabatini BL, Svoboda K. Structure and function of dendritic spines. Annu Rev Physiol. 2002;64:313–53.
36. Wilhelm BG, Mandad S, Truckenbrodt S, Kröhnert K, Schäfer C, Rammner B, Koo SJ, Claßen GA, Krauss M, Haucke V, Urlaub H, Rizzoli SO. Composition of isolated synaptic boutons reveals the amounts of vesicle trafficking proteins. Science. 2014;344(6187):1023–8.
37. Harris JJ, Jolivet R, Attwell D. Synaptic energy use and supply. Neuron. 2012;75(5):762–77.
38. Wang YC, Lauwers E, Verstreken P. Presynaptic protein homeostasis and neuronal function. Curr Opin Genet Dev. 2017;44:38–46.
39. Liang Y, Sigrist S. Autophagy and proteostasis in the control of synapse aging and disease. Curr Opin Neurobiol. 2018;48:113–21.
40. Eriksson PS, Perfilieva E, Björk-Eriksson T, Alborn A-M, Nordborg C, Peterson DA, Gage FH. Neurogenesis in the adult human hippocampus. Nat Med. 1998;4:1313.

41. Sorrells SF, Paredes MF, Cebrian-Silla A, Sandoval K, Qi D, Kelley KW, James D, Mayer S, Chang J, Auguste KI, Chang EF, Gutierrez AJ, Kriegstein AR, Mathern GW, Oldham MC, Huang EJ, Garcia-Verdugo JM, Yang Z, Alvarez-Buylla A. Human hippocampal neurogenesis drops sharply in children to undetectable levels in adults. Nature. 2018;555(7696):377–81.
42. Kaushik S, Cuervo AM. Proteostasis and aging. Nat Med. 2015;21(12):1406–15.
43. Mizushima N, Komatsu M. Autophagy: renovation of cells and tissues. Cell. 2011;147(4):728–41.
44. Mijaljica D, Prescott M, Devenish RJ. Microautophagy in mammalian cells: revisiting a 40-year-old conundrum. Autophagy. 2014;7(7):673–82.
45. Kaushik S, Cuervo AM. Chaperone-mediated autophagy: a unique way to enter the lysosome world. Trends Cell Biol. 2012;22(8):407–17.
46. Stolz A, Ernst A, Dikic I. Cargo recognition and trafficking in selective autophagy. Nat Cell Biol. 2014;16(6):495–501.
47. Jiang P, Mizushima N. Autophagy and human diseases. Cell Res. 2014;24(1):69–79.
48. Choi AMK, Ryter SW, Levine B. Autophagy in human health and disease. N Engl J Med. 2013;368(7):651–62.
49. Kroemer G. Autophagy: a druggable process that is deregulated in aging and human disease. J Clin Invest. 2015;125(1):1–4.
50. Hara T, Nakamura K, Matsui M, Yamamoto A, Nakahara Y, Suzuki-Migishima R, Yokoyama M, Mishima K, Saito I, Okano H, Mizushima N. Suppression of basal autophagy in neural cells causes neurodegenerative disease in mice. Nature. 2006;441(7095):885–9.
51. Komatsu M, Waguri S, Chiba T, Murata S, Iwata J, Tanida I, Ueno T, Koike M, Uchiyama Y, Kominami E, Tanaka K. Loss of autophagy in the central nervous system causes neurodegeneration in mice. Nature. 2006;441(7095):880–4.
52. Fimia GM, Stoykova A, Romagnoli A, Giunta L, Di Bartolomeo S, Nardacci R, Corazzari M, Fuoco C, Ucar A, Schwartz P, Gruss P, Piacentini M, Chowdhury K, Cecconi F. Ambra1 regulates autophagy and development of the nervous system. Nature. 2007;447(7148):1121–5.
53. Dragich JM, Kuwajima T, Hirose-Ikeda M, Yoon MS, Eenjes E, Bosco JR, Fox LM, Lystad AH, Oo TF, Yarygina O, Mita T, Waguri S, Ichimura Y, Komatsu M, Simonsen A, Burke RE, Mason CA, Yamamoto A. Autophagy linked FYVE (Alfy/WDFY3) is required for establishing neuronal connectivity in the mammalian brain. Elife. 2016;5:e14810.
54. Tang G, Gudsnuk K, Kuo SH, Cotrina ML, Rosoklija G, Sosunov A, Sonders MS, Kanter E, Castagna C, Yamamoto A, Yue Z, Arancio O, Peterson BS, Champagne F, Dwork AJ, Goldman J, Sulzer D. Loss of mTOR-dependent macroautophagy causes autistic-like synaptic pruning deficits. Neuron. 2014;83(5):1131–43.
55. Yoshii SR, Kuma A, Akashi T, Hara T, Yamamoto A, Kurikawa Y, Itakura E, Tsukamoto S, Shitara H, Eishi Y, Mizushima N. Systemic analysis of *Atg5*-Null mice rescued from neonatal lethality by transgenic ATG5 expression in neurons. Dev Cell. 2016;39(1):116–30.
56. Nixon RA, Wegiel J, Kumar A, Yu WH, Peterhoff C, Cataldo A, Cuervo AM. Extensive involvement of autophagy in Alzheimer disease: an immuno-electron microscopy study. J Neuropathol Exp Neurol. 2005;64(2):113–22.
57. Yamamoto A, Yue Z. Autophagy and its normal and pathogenic states in the brain. Annu Rev Neurosci. 2014;37:55–78.
58. Menzies FM, Fleming A, Caricasole A, Bento CF, Andrews SP, Ashkenazi A, Fullgrabe J, Jackson A, Jimenez Sanchez M, Karabiyik C, Licitra F, Lopez Ramirez A, Pavel M, Puri C, Renna M, Ricketts T, Schlotawa L, Vicinanza M, Won H, Zhu Y, Skidmore J, Rubinsztein DC. Autophagy and neurodegeneration: pathogenic mechanisms and therapeutic opportunities. Neuron. 2017;93(5):1015–34.
59. Kim M, Ho A, Lee JH. Autophagy and human neurodegenerative diseases-a Fly's perspective. Int J Mol Sci. 2017;18(7):pii: E1596.
60. Ramesh N, Pandey UB. Autophagy dysregulation in ALS: when protein aggregates get out of hand. Front Mol Neurosci. 2017;10:263.

61. Hars ES, Qi H, Jin SV, Cai L, Hu C, Liu LF. Autophagy regulates ageing in *C. elegans*. Autophagy. 2014;3(2):93–5.
62. Gupta VK, Scheunemann L, Eisenberg T, Mertel S, Bhukel A, Koemans TS, Kramer JM, Liu KS, Schroeder S, Stunnenberg HG, Sinner F, Magnes C, Pieber TR, Dipt S, Fiala A, Schenck A, Schwaerzel M, Madeo F, Sigrist SJ. Restoring polyamines protects from age-induced memory impairment in an autophagy-dependent manner. Nat Neurosci. 2013;16(10):1453–60.
63. Lin FH, Zhang WL, Li H, Tian XD, Zhang J, Li X, Li CY, Tan JH. Role of autophagy in modulating post-maturation aging of mouse oocytes. Cell Death Dis. 2018;9(3):308.
64. Yang F, Chu X, Yin M, Liu X, Yuan H, Niu Y, Fu L. mTOR and autophagy in normal brain aging and caloric restriction ameliorating age-related cognition deficits. Behav Brain Res. 2014;264:82–90.
65. Perluigi M, Di Domenico F, Butterfield DA. mTOR signaling in aging and neurodegeneration: at the crossroad between metabolism dysfunction and impairment of autophagy. Neurobiol Dis. 2015;84:39–49.
66. Yu Y, Feng L, Li J, Lan X, Lixiang A, Lv X, Zhang M, Chen L. The alteration of autophagy and apoptosis in the hippocampus of rats with natural aging-dependent cognitive deficits. Behav Brain Res. 2017;334:155–62.
67. Lipinski MM, Zheng B, Lu T, Yan Z, Py BF, Ng A, Xavier RJ, Li C, Yankner BA, Scherzer CR, Yuan J. Genome-wide analysis reveals mechanisms modulating autophagy in normal brain aging and in Alzheimer's disease. Proc Natl Acad Sci. 2010;107(32):14164–9.
68. Alirezaei M, Kiosses WB, Flynn CT, Brady NR, Fox HS. Disruption of neuronal autophagy by infected microglia results in neurodegeneration. PLoS One. 2008;3(8):e2906.
69. Gan L, Vargas MR, Johnson DA, Johnson JA. Astrocyte-specific overexpression of Nrf2 delays motor pathology and synuclein aggregation throughout the CNS in the alpha-synuclein mutant (A53T) mouse model. J Neurosci. 2012;32(49):17775–87.
70. Madill M, McDonagh K, Ma J, Vajda A, McLoughlin P, O'Brien T, Hardiman O, Shen S. Amyotrophic lateral sclerosis patient iPSC-derived astrocytes impair autophagy via non-cell autonomous mechanisms. Mol Brain. 2017;10(1):22.
71. Davis CH, Kim KY, Bushong EA, Mills EA, Boassa D, Shih T, Kinebuchi M, Phan S, Zhou Y, Bihlmeyer NA, Nguyen JV, Jin Y, Ellisman MH, Marsh-Armstrong N. Transcellular degradation of axonal mitochondria. Proc Natl Acad Sci U S A. 2014;111(26):9633–8.
72. Melentijevic I, Toth ML, Arnold ML, Guasp RJ, Harinath G, Nguyen KC, Taub D, Parker JA, Neri C, Gabel CV, Hall DH, Driscoll M. *C. elegans* neurons jettison protein aggregates and mitochondria under neurotoxic stress. Nature. 2017;542(7641):367–71.
73. Kulkarni A, Chen J, Maday S. Neuronal autophagy and intercellular regulation of homeostasis in the brain. Curr Opin Neurobiol. 2018;51:29–36.
74. Simonsen A, Cumming RC, Brech A, Isakson P, Schubert DR, Finley KD. Promoting basal levels of autophagy in the nervous system enhances longevity and oxidant resistance in adult *Drosophila*. Autophagy. 2014;4(2):176–84.
75. Juhasz G, Erdi B, Sass M, Neufeld TP. Atg7-dependent autophagy promotes neuronal health, stress tolerance, and longevity but is dispensable for metamorphosis in *Drosophila*. Genes Dev. 2007;21(23):3061–6.
76. de Cabo R, Carmona-Gutierrez D, Bernier M, Hall MN, Madeo F. The search for antiaging interventions: from elixirs to fasting regimens. Cell. 2014;157(7):1515–26.
77. Luo L, Dai JR, Guo SS, Lu AM, Gao XF, Gu YR, Zhang XF, Xu HD, Wang Y, Zhu Z, Wood LJ, Qin ZH. Lysosomal proteolysis is associated with exercise-induced improvement of mitochondrial quality control in aged hippocampus. J Gerontol A Biol Sci Med Sci. 2017;72:1342.
78. Bjedov I, Toivonen JM, Kerr F, Slack C, Jacobson J, Foley A, Partridge L. Mechanisms of life span extension by rapamycin in the fruit fly *Drosophila melanogaster*. Cell Metab. 2010;11(1):35–46.
79. Harrison DE, Strong R, Sharp ZD, Nelson JF, Astle CM, Flurkey K, Nadon NL, Wilkinson JE, Frenkel K, Carter CS, Pahor M, Javors MA, Fernandez E, Miller RA. Rapamycin fed late in life extends lifespan in genetically heterogeneous mice. Nature. 2009;460(7253):392–5.

80. King MA, Hands S, Hafiz F, Mizushima N, Tolkovsky AM, Wyttenbach A. Rapamycin inhibits polyglutamine aggregation independently of autophagy by reducing protein synthesis. Mol Pharmacol. 2008;73(4):1052–63.
81. Wyttenbach A, Hands S, King MA, Lipkow K, Tolkovsky AM. Amelioration of protein misfolding disease by rapamycin: translation or autophagy? Autophagy. 2008;4(4):542–5.
82. Halloran J, Hussong SA, Burbank R, Podlutskaya N, Fischer KE, Sloane LB, Austad SN, Strong R, Richardson A, Hart MJ, Galvan V. Chronic inhibition of mammalian target of rapamycin by rapamycin modulates cognitive and non-cognitive components of behavior throughout lifespan in mice. Neuroscience. 2012;223:102–13.
83. Li J, Kim SG, Blenis J. Rapamycin: one drug, many effects. Cell Metab. 2014;19(3):373–9.
84. Eisenberg T, Knauer H, Schauer A, Buttner S, Ruckenstuhl C, Carmona-Gutierrez D, Ring J, Schroeder S, Magnes C, Antonacci L, Fussi H, Deszcz L, Hartl R, Schraml E, Criollo A, Megalou E, Weiskopf D, Laun P, Heeren G, Breitenbach M, Grubeck-Loebenstein B, Herker E, Fahrenkrog B, Frohlich KU, Sinner F, Tavernarakis N, Minois N, Kroemer G, Madeo F. Induction of autophagy by spermidine promotes longevity. Nat Cell Biol. 2009;11(11):1305–14.
85. Madeo F, Eisenberg T, Büttner S, Ruckenstuhl C, Kroemer G. Spermidine: a novel autophagy inducer and longevity elixir. Autophagy. 2010;6(1):160–2.
86. Morselli E, Marino G, Bennetzen MV, Eisenberg T, Megalou E, Schroeder S, Cabrera S, Benit P, Rustin P, Criollo A, Kepp O, Galluzzi L, Shen S, Malik SA, Maiuri MC, Horio Y, Lopez-Otin C, Andersen JS, Tavernarakis N, Madeo F, Kroemer G. Spermidine and resveratrol induce autophagy by distinct pathways converging on the acetylproteome. J Cell Biol. 2011;192(4):615–29.
87. Lubas M, Harder LM, Kumsta C, Tiessen I, Hansen M, Andersen JS, Lund AH, Frankel LB. eIF5A is required for autophagy by mediating ATG3 translation. EMBO Rep. 2018;19(6):pii: e46072.
88. Wolff EC, Kang KR, Kim YS, Park MH. Posttranslational synthesis of hypusine: evolutionary progression and specificity of the hypusine modification. Amino Acids. 2007;33(2):341–50.
89. Margulies C, Tully T, Dubnau J. Deconstructing memory in *Drosophila*. Curr Biol. 2005;15(17):R700–13.
90. Saitoe M, Horiuchi J, Tamura T, Ito N. *Drosophila* as a novel animal model for studying the genetics of age-related memory impairment. Rev Neurosci. 2005;16(2):137–50.
91. Tonoki A, Davis RL. Aging impairs intermediate-term behavioral memory by disrupting the dorsal paired medial neuron memory trace. Proc Natl Acad Sci U S A. 2012;109(16):6319–24.
92. Murakami S, Minami-Ohtsubo M, Nakato R, Shirahige K, Tabata T. Two components of aversive memory in *Drosophila*, anesthesia-sensitive and anesthesia-resistant memory, require distinct domains within the Rgk1 small GTPase. J Neurosci. 2017;37(22):5496–510.
93. Tamura T, Chiang A-S, Ito N, Liu H-P, Horiuchi J, Tully T, Saitoe M. Aging specifically impairs amnesiac-dependent memory in *Drosophila*. Neuron. 2003;40(5):1003–11.
94. Takeuchi T, Duszkiewicz AJ, Morris RG. The synaptic plasticity and memory hypothesis: encoding, storage and persistence. Philos Trans R Soc Lond B Biol Sci. 2014;369(1633):20130288.
95. Petzoldt AG, Lutzkendorf J, Sigrist SJ. Mechanisms controlling assembly and plasticity of presynaptic active zone scaffolds. Curr Opin Neurobiol. 2016;39:69–76.
96. Gupta VK, Pech U, Bhukel A, Fulterer A, Ender A, Mauermann SF, Andlauer TF, Antwi-Adjei E, Beuschel C, Thriene K, Maglione M, Quentin C, Bushow R, Schwarzel M, Mielke T, Madeo F, Dengjel J, Fiala A, Sigrist SJ. Spermidine suppresses age-associated memory impairment by preventing adverse increase of presynaptic active zone size and release. PLoS Biol. 2016;14(9):e1002563.
97. Bhukel A, Madeo F, Sigrist SJ. Spermidine boosts autophagy to protect from synapse aging. Autophagy. 2017;13(2):444–5.
98. Soukup SF, Kuenen S, Vanhauwaert R, Manetsberger J, Hernandez-Diaz S, Swerts J, Schoovaerts N, Vilain S, Gounko NV, Vints K, Geens A, De Strooper B, Verstreken P. A

LRRK2-dependent EndophilinA phosphoswitch is critical for macroautophagy at presynaptic terminals. Neuron. 2016;92(4):829–44.
99. Vanhauwaert R, Kuenen S, Masius R, Bademosi A, Manetsberger J, Schoovaerts N, Bounti L, Gontcharenko S, Swerts J, Vilain S, Picillo M, Barone P, Munshi ST, de Vrij FM, Kushner SA, Gounko NV, Mandemakers W, Bonifati V, Meunier FA, Soukup SF, Verstreken P. The SAC1 domain in synaptojanin is required for autophagosome maturation at presynaptic terminals. EMBO J. 2017;36:1392.
100. Okerlund ND, Schneider K, Leal-Ortiz S, Montenegro-Venegas C, Kim SA, Garner LC, Gundelfinger ED, Reimer RJ, Garner CC. Bassoon controls presynaptic autophagy through Atg5. Neuron. 2017;93(4):897–913.e897.
101. Binotti B, Pavlos NJ, Riedel D, Wenzel D, Vorbruggen G, Schalk AM, Kuhnel K, Boyken J, Erck C, Martens H, Chua JJ, Jahn R. The GTPase Rab26 links synaptic vesicles to the autophagy pathway. Elife. 2015;4:e05597.
102. Wucherpfennig T, Wilsch-Brauninger M, Gonzalez-Gaitan M. Role of *Drosophila* Rab5 during endosomal trafficking at the synapse and evoked neurotransmitter release. J Cell Biol. 2003;161(3):609–24.
103. Uytterhoeven V, Kuenen S, Kasprowicz J, Miskiewicz K, Verstreken P. Loss of skywalker reveals synaptic endosomes as sorting stations for synaptic vesicle proteins. Cell. 2011;145(1):117–32.
104. Rowland AM, Richmond JE, Olsen JG, Hall DH, Bamber BA. Presynaptic terminals independently regulate synaptic clustering and autophagy of GABAA receptors in *Caenorhabditis elegans*. J Neurosci. 2006;26(6):1711–20.
105. Shehata M, Matsumura H, Okubo-Suzuki R, Ohkawa N, Inokuchi K. Neuronal stimulation induces autophagy in hippocampal neurons that is involved in AMPA receptor degradation after chemical long-term depression. J Neurosci. 2012;32(30):10413–22.
106. Nikoletopoulou V, Sidiropoulou K, Kallergi E, Dalezios Y, Tavernarakis N. Modulation of autophagy by BDNF underlies synaptic plasticity. Cell Metab. 2017;26(1):230–242.e235.
107. Narendra D, Tanaka A, Suen DF, Youle RJ. Parkin is recruited selectively to impaired mitochondria and promotes their autophagy. J Cell Biol. 2008;183(5):795–803.
108. Narendra DP, Jin SM, Tanaka A, Suen DF, Gautier CA, Shen J, Cookson MR, Youle RJ. PINK1 is selectively stabilized on impaired mitochondria to activate Parkin. PLoS Biol. 2010;8(1):e1000298.
109. Cai Q, Zakaria HM, Simone A, Sheng ZH. Spatial parkin translocation and degradation of damaged mitochondria via mitophagy in live cortical neurons. Curr Biol. 2012;22(6):545–52.
110. Sung H, Tandarich LC, Nguyen K, Hollenbeck PJ. Compartmentalized regulation of Parkin-mediated mitochondrial quality control in the *Drosophila* nervous system in vivo. J Neurosci. 2016;36(28):7375–91.
111. Ashrafi G, Schlehe JS, LaVoie MJ, Schwarz TL. Mitophagy of damaged mitochondria occurs locally in distal neuronal axons and requires PINK1 and Parkin. J Cell Biol. 2014;206(5):655–70.
112. Cao X, Wang H, Wang Z, Wang Q, Zhang S, Deng Y, Fang Y. In vivo imaging reveals mitophagy independence in the maintenance of axonal mitochondria during normal aging. Aging Cell. 2017;16(5):1180–90.
113. Pickles S, Vigie P, Youle RJ. Mitophagy and quality control mechanisms in mitochondrial maintenance. Curr Biol. 2018;28(4):R170–85.
114. Maday S, Wallace KE, Holzbaur EL. Autophagosomes initiate distally and mature during transport toward the cell soma in primary neurons. J Cell Biol. 2012;196(4):407–17.
115. Wang T, Martin S, Papadopulos A, Harper CB, Mavlyutov TA, Niranjan D, Glass NR, Cooper-White JJ, Sibarita JB, Choquet D, Davletov B, Meunier FA. Control of autophagosome axonal retrograde flux by presynaptic activity unveiled using botulinum neurotoxin type a. J Neurosci. 2015;35(15):6179–94.
116. Stavoe AK, Hill SE, Hall DH, Colon-Ramos DA. KIF1A/UNC-104 transports ATG-9 to regulate neurodevelopment and autophagy at synapses. Dev Cell. 2016;38(2):171–85.

117. Mizushima N, Yamamoto A, Matsui M, Yoshimori T, Ohsumi Y. In vivo analysis of autophagy in response to nutrient starvation using transgenic mice expressing a fluorescent autophagosome marker. Mol Biol Cell. 2004;15(3):1101–11.
118. Maday S, Holzbaur EL. Compartment-specific regulation of autophagy in primary neurons. J Neurosci. 2016;36(22):5933–45.
119. Tsvetkov AS, Miller J, Arrasate M, Wong JS, Pleiss MA, Finkbeiner S. A small-molecule scaffold induces autophagy in primary neurons and protects against toxicity in a Huntington disease model. Proc Natl Acad Sci U S A. 2010;107(39):16982–7.
120. Fox JH, Connor T, Chopra V, Dorsey K, Kama JA, Bleckmann D, Betschart C, Hoyer D, Frentzel S, Difiglia M, Paganetti P, Hersch SM. The mTOR kinase inhibitor Everolimus decreases S6 kinase phosphorylation but fails to reduce mutant huntingtin levels in brain and is not neuroprotective in the R6/2 mouse model of Huntington's disease. Mol Neurodegener. 2010;5:26.
121. Shen W, Ganetzky B. Autophagy promotes synapse development in *Drosophila.* J Cell Biol. 2009;187(1):71–9.
122. Kononenko NL, Classen GA, Kuijpers M, Puchkov D, Maritzen T, Tempes A, Malik AR, Skalecka A, Bera S, Jaworski J, Haucke V. Retrograde transport of TrkB-containing autophagosomes via the adaptor AP-2 mediates neuronal complexity and prevents neurodegeneration. Nat Commun. 2017;8:14819.
123. Burk K, Murdoch JD, Freytag S, Koenig M, Bharat V, Markworth R, Burkhardt S, Fischer A, Dean C. EndophilinAs regulate endosomal sorting of BDNF-TrkB to mediate survival signaling in hippocampal neurons. Sci Rep. 2017;7(1):2149.
124. Wu CL, Chen CH, Hwang CS, Chen SD, Hwang WC, Yang DI. Roles of p62 in BDNF-dependent autophagy suppression and neuroprotection against mitochondrial dysfunction in rat cortical neurons. J Neurochem. 2017;140(6):845–61.

# Chapter 10
# Virus and Autophagy: Enemies or Allies

**José A. Boga, Zulema Pérez-Martínez, Ana Coto-Montes, Marta E. Alvarez-Argüelles, and Russel J. Reiter**

**Abstract** Viral infection is one of the several stimuli which trigger autophagy, a self-degradative process that is important for balancing sources of energy at critical times in development and in response to nutrient stress. This process also plays a housekeeping role in removing misfolded or aggregated proteins and clearing damaged organelles. Virus-induced autophagy has a dual role since it may be beneficial to the host by eliminating intracellular pathogens or it may benefit some viruses, which have developed strategies to directly or indirectly subvert autophagy in order to promote different stages of the viral life cycle. The upregulation of both oxidative and endoplasmic reticulum stresses has been reported as a means by which virus-induced pathways trigger autophagy. In this chapter the relationships between autophagy and viral infection are considered.

## 10.1 Introduction

Autophagy is a self-degradative process which, under basal conditions, allows cells to break down long-lived proteins and damaged organelles. It can be induced under conditions of cellular stress, such as starvation, growth factor deprivation, hypoxia, oxidation of critical molecules, DNA damage, protein aggregation, or infection by intracellular pathogens. Autophagy, which is important for balancing sources of

J. A. Boga (✉) · Z. Pérez-Martínez · M. E. Alvarez-Argüelles
Servicio de Microbiología, Hospital Universitario Central de Asturias, Oviedo, Spain

Instituto de Investigación Sanitaria del Principado de Asturias, ISPA, Oviedo, Spain
e-mail: joseantonio.boga@sespa.es

A. Coto-Montes
Departamento de Morfología y Biología Celular, Universidad de Oviedo, Oviedo, Spain

Instituto de Investigación Sanitaria del Principado de Asturias, ISPA, Oviedo, Spain

R. J. Reiter
Department of Cellular & Structural Biology, UT Health Science Center at San Antonio, San Antonio, TX, USA



K. Turksen (ed.), *Autophagy in Health and Disease*, Stem Cell Biology and Regenerative Medicine, https://doi.org/10.1007/978-3-319-98146-8_10

energy at critical times in development and in response to nutrient stress, exhibits a dual role, facilitating stress adaptation and promoting cell survival or providing an alternative pathway of cell death [1]. In the case of viral infection-induced autophagy, it may be part of the defense response of the host or promote viral replication at different stages.

## 10.2 Autophagy Molecular Pathway

Autophagy is a complex process that can be categorized into several sequential steps at molecular and cellular levels. The autophagic machinery requires the formation of a double membrane structure, the phagophore, which is generated de novo from the endoplasmic reticulum, Golgi apparatus, or other plasma membrane-derived endocytic organelles. This involves the mediation of ULK1/FIP200/ATG101/ATG13 protein kinase and VPS34/beclin1/VPS15/ATG14 lipid kinase complexes. These complexes induce the formation of the ATG5/ATG12/ATG16L1 complex (formed with the help of ATG7 and ATG10), which promotes the elongation of the phagophore the latter eventually engulfs a portion of the cytosol or specific cargoes (proteins, lipids, organelles), forming a double-membraned vacuole called the autophagosome. Aided by ATG3 and ATG7, these complexes facilitate the addition of a phosphatidylethanolamine group to the cytosolic form of mammalian LC3 homologues (LC3A, LC3B, LC3C, Gabarap, Gabarap-L1, and Gabarap-L2), which is referred to as the LC3-I to LC3-II conversion. The lipid-bound form of LC3 homologues is then recruited to the autophagosome, which matures and fuses with the lysosome, forming a single-membraned vesicle termed the autolysosome. The contents are then degraded by several lysosomal membrane proteins (GTPase RAB7, LAMP1, LAMP2, as well as SNARE proteins, such as syntaxin 17 and SNAP29) with the aid of another beclin1/VPS34 complex, where the UV radiation resistance-associated gene (UVRAG), rather than ATG14, is required [2].

## 10.3 Autophagy and Virus Infection

The induction of autophagy is triggered by a variety of stressful stimuli, including nutrient deprivation, danger-associated molecular patterns (DAMPs), ER stress, hypoxia, redox stress and mitochondrial damage. Many of these stimuli can be triggered by viruses during different stages of their replication cycle [3]. Thus, a cellular process dedicated to degrading cytosolic contents that also engulfs and destroys pathogens has been described and termed "xenophagy" [4]. The fact that virus-induced autophagy is capable of preventing the early apoptotic death of cells suggests that xenophagy might limit the cytopathic effect of viruses and the pathological consequences associated with cell death triggered by viral infection [5].

One autophagic stimulus is the attachment of the virus particle to the host cell. Adenovirus types B and D, human herpesvirus 6, and measles virus stimulate autophagy upon viral binding to the cell surface receptor CD46. In case of the measles virus, autophagy is activated by Vps34-Beclin-1 complex via interaction of CD46 with Golgi-associated PDZ and coiled-coil motif-containing protein (GOPC) [6]. The binding of the C-terminal domain of the fusogenic gp41 subunit of HIV to $CD4^{+}$ T-cells triggers autophagy [7, 8].

The first mechanism to limit the extent of viral spread involves a large repertoire of pattern recognition receptors (PRRs), including Toll-like receptors (TLRs), Nod-like receptors (NLRs), RIG-I-like receptors (RLRs), and AIM2-like receptors (ALRs). These receptors recognize pathogen-associated molecular patterns (PAMPs), which are mostly viral nucleic acids or their synthetic analogs produced during the viral infection [9–12]. As an organelle gathering cytosolic content into a double-membrane vesicle that fuses with endosomal compartments, the autophagosome can specifically deliver intracellular PAMPs to endosomal PRRs and MHC-loading compartments to initiate innate and adaptive immune responses. Thus, Coxsackie virus B3 (CVB3)-infected kidney fibroblasts show autophagy-dependent activation of TLR3 [13]. Vesicular stomatitis virus (VSV)-infected plasmacytoid dendritic cells (pDCs) produce interferon-α (IFN-α); this production is dependent on the autophagic delivery of viral replication intermediates to TLR7 present on late endosomes. The fact that ATG5−/− VSV-infected mice exhibited higher viral loads, suggests a control of VSV replication in vivo by autophagy [14].

Induction of autophagy also occurs via direct activation of TLRs [15]. Upon stimulation of TLR 3, 4 and 7 with different purified PAMPs, myeloid differentiation primary response gene 88 (MyD88) and TIR-domain-containing adapter-inducing interferon-β (TRIF) interact with beclin-1 leading to the formation of beclin-1-Vsp34 complexes that induce autophagy [16, 17]. The role of autophagy in delivering viral PAMPs to TLRs and the subsequent induction of autophagy upon TLR activation play important roles in the autophagic sequestration and delivery of PAMPs to endosomal compartments. They also serve for autolysosomal degradation of virions and viral proteins, as well as for antigen presentation via MHC-I and MHC-II molecules, thereby triggering the adaptive immune response.

The cytoplasm offers a diverse range of autophagic targets that vary in size and complexity, ranging from protein aggregates up to complete organelles, which can be selectively recognized and sequestered by proteins that function as autophagic adaptors. The main autophagic adaptors that are classified as sequestosome-1/p62-like receptors (SLR) in response to bacterial and viral infections include p62, neighbor of BRCA1 gene 1 (NBR1), nuclear dot protein of Mr 52,000 (NDP52) and optineurin. Cargo receptors typically have one or more LC3-interacting region (LIR) and a cargo-binding domain such as an ubiquitin-associated (UBA) domain for ubiquitylated proteins, as well as additional protein interaction domains that are involved in inflammatory processes [18]. Autophagic degradation of Sindbis virus, which involves the selective degradation of the viral capsid in a p62-dependent manner, is an example of this mechanism [19].

Viral replication itself frequently elicits stress responses, such as production of reactive oxygen species (ROS) or ER stress that induce autophagy. ER stress, which is produced by an accumulation of misfolded or unfolded proteins in this compartment, activates an unfolded protein response (UPR) mediated by three ER membrane-associated proteins. These three proteins, PERK (PKR-like eIF2α kinase) ATF6 (activating transcription factor 6) and IRE1 (inositol requiring enzyme 1), are normally bound to and inactivated by the chaperone BiP/GRP78 at the side of the ER lumen. The BiP interaction with unfolded luminal proteins leads the release of PERK, IRE1 and ATF6. Among other effects, it has been reported that while PERK and ATF6 act as inducers of autophagy, IRE1 functions as a negative regulator [20]. PERK mediates the phosphorylation of eukaryotic translation initiation factor 2α (eIF2α). This results in rapid reduction of mRNA translation, thereby reducing the load of new proteins in the ER. This also leads to the selective translation of activating transcription factor 4 (ATF4) and C/EBP homologous protein (CHOP) to induce the transcriptional activation of the proteins LC3 and Atg5. In a recent study, sets of autophagic genes including ATG3, ATG12, ATG16, Map, Beclin1, and Gabarapl2 were identified as targets for ATF4. Meanwhile, transcription of autophagic genes including Sqstm1, NBR1, and ATG7 were boosted when ATF4 and CHOP were coactivated [21]. IRE1 is responsible for the unconventional splicing of the X box-binding protein 1 (XBP1) mRNA, which regulates transcription of genes encoding ER chaperones, biogenesis of phospholipids and ER-associated protein degradation (ERAD) components [22] Moreover, it has been hypothesized that IRE1/XBP1-dependent signals dampen excessive autophagy triggered via the PERK/eIF2α pathway [21]. The cytoplasmic domain of ATF6, which is released by regulated intramembrane proteolysis (RIP) by site 1 and site 2 proteases, functions as a transcription factor to transactivate genes encoding ER chaperones, components of the ERAD machinery and protein foldases. ATF6 interacts with transcription factor C/EBP-β and induce expression of *Dapk1*, which promotes apoptosis and autophagy and suppresses cell division [23].

HCV infection induces ER stress and activates all three sensors of UPR [20]. The silencing of any of the three UPR signaling proteins, mediated by siRNA, results in a significant reduction of the amount of LC3-II and an inhibition in HCV replication [23]. On the other hand, subgenomic replicons expressing nonstructural proteins (NS3-5B), as well as a mutant virus lacking the envelope glycoproteins, potently induced autophagy in the absence of detectable UPR [24]. Nevertheless, expression of the replicase protein NS4A of other related flaviviruses, such as dengue virus (DENV) and Modoc virus induce autophagy via UPR [25]. DNA viruses, such as VZV, can also induce autophagy via ER-stress and UPR-related pathways. Abundant VZV glycoprotein biosynthesis induces ER stress, which in turn triggers UPR and thus autophagy to maintain cellular homeostasis [26].

Autophagic process are also associated with changes in the cellular production of ROS and reactive nitrogen species (RNS), a range of oxygen- and nitrogen-derived molecules formed during oxidative metabolism. Among them, $O_2^{\bullet-}$ and $H_2O_2$, which can be formed by controlled mechanisms in cells and act as cell signaling molecules, interact with NO to generate nitrating species, such as $ONOO^-$. ROS and $ONOO^-$ oxidize lipids to produce reactive lipid species (RLS). ROS and RNS can modify proteins posttranslationally and, thus regulate antioxidant defence and autophagy in the cell. A prominent redox signaling pathway responsive to ROS/RNS/RLS is nuclear factor-erythroid 2-related factor 2/Kelch-like ECH (enoyl-CoA hydratase)-associated protein 1 (Nrf2/Keap1) pathway. Nrf2 can bind to the ARE (antioxidant-response element) sequence to activate transcription of antioxidant and cellular detoxification genes. Nrf2 is negatively regulated by the cytoplasmic redox-sensor protein Keap1 via its Nrf2/ECH homology 2 (Neh2) domain. The modification of a cysteine residue of Keap1 release Nrf2, which is translocated to the nucleus. The released Nrf2 activates the transcription of both antioxidant and autophagic genes, such as p62 [27]. Thus, autophagy can be indirectly regulated by transcriptional mechanisms in response to ROS/RNS via Nrf2/Keap1 pathway. The role of Nrf2 in viral infections is currently under investigation. Nrf2-mediated antioxidant pathways play a pivotal role in protection against the oxidant-induced exacerbation of influenza A virus infection by the upregulation of antioxidants [28]. DENV infection results in an intracellular accumulation of ROS. A role for the transcription factor Nrf2 in limiting both antiviral and cell death responses to the virus by feedback modulation of oxidative stress has been proposed [29]. An induction in the expression of Nrf2-regulated cytoprotective genes by HBV has been reported, suggesting that an increase in Nrf2-dependent antioxidants could protect HBV-infected cells from oxidative-mediated cell damage [30].

Autophagy also plays a key role in the adaptive immune response by processing and delivering antigens for presentation in complex with MHC-I and MHC-II molecules [31]. In the case of EBV, the viral protein EBNA1 is targeted to MHC-II compartments via autophagic uptake and fusion of autophagosomes with these compartments [32]. Blocking of autophagy specifically inhibits endogenous MHC class II processing of EBNA1 [33]. While autophagy is not required for viral antigen processing and presentation of MHC-I peptides at early time points HSV-1 after infection, it comes into play at a late stage of infection [34].

Despite autophagy can play a key role in antiviral defense, such as has been shown below, autophagy or individual factors of the autophagy pathway can also enhance viral replication. Thus, viruses have developed strategies to directly or indirectly subvert autophagy in order to promote different stages of the viral life cycle.

Most positive-strand RNA viruses trigger autophagy to reshape the endomembrane system in order to create membrane-associated replication factories [35]. Double-membrane vesicles (DMVs), which are characteristic of autophagosome-like structures, were detected in cells infected by several picornaviruses, such as poliovirus, coxsackievirus B3 (CVB3), enterovirus 71 (EV71), and foot-and-mouth disease virus (FMDV). Moreover, the formation of autophagosome-like DMVs is induced by the lipidation of LC3, which is mediated by expression of the picornaviral proteins 2BC and 3A [36]. Although autophagosomes, per se, are dispensable for the biogenesis of viral replication sites, in some cases a decrease in virus production is observed in the absence of autophagy [37, 38].

Autophagy can also promote viral replication. During EBV replication, autophagy is blocked at the final degradative steps. Because of the blockade, EBV hijacks the autophagic vesicles for its intracellular transportation and enhances viral replication allowing the virus to skip its degradation by the lysosomes [39]. In addition, an essential viral protein for Kaposi's sarcoma-associated herpesvirus (KSHV) lytic reactivation, the replication and transcription activator (RTA), enhances the autophagic process, leading to an elevation in the number of autophagic vacuoles, an increase in the level of the lipid bound form of LC3 protein, and the formation of autolysosomes. Moreover, the inhibition of autophagy affects RTA-mediated lytic gene expression and viral DNA replication [40]. Immunity-associated GTPase family M (*IRGM*) is a GTP-binding protein with a key function in the innate immune response by regulating autophagy induction in response to intracellular pathogens. IRGM, which interacts with the autophagy proteins Atg5, Atg10, *microtubule*-associated proteins 1A/1B light chain 3 beta 2 (MAP1CL3C) and SH3-domain GRB2-like endophilin B1 (SH3GLB1), has been recently shown to be a common target of several RNA viruses which subvert the autophagy network. Silencing of IRGM expression impaired both measles virus (MeV), HCV and HIV-1 induced autophagy and viral replication [41]. Japanese encephalitis virus (JEV) infection-induced autophagy can promote viral replication. Cells transfected with a plasmid that expresses the viral protein NS3 show that NS3 signals are completely colocalized with the IRGM signals, documenting that NS3 could target IRGM which may play a role in the replication of JEV [42]

Upon replication of the viral genome, it must be packaged into a virus particle that is released from the infected cell. For many viruses, this particle is surrounded by a double membrane envelope. Autophagosomes might serve as a membrane source. Thus, efficient envelopment of HBV was shown to depend on induction of autophagy. Furthermore, the fact that major HBV envelope protein (HBsAg) binds to and colocalizes with LC3-I and LC3-II during HBV infection indicates that this interaction might be important for acquiring the viral envelope [43]. On the other hand, autophagy has been shown to serve also in the nonlytic release of nonenveloped viruses via the formation of DMVs, such as those described in the case of picornaviruses (poliovirus, rhinovirus 2, or rhinovirus 14). A potential mechanism for the release of cytosolic picornaviral particles via the formation of these DMVs is supported by the presence of both LC3 and poliovirus capsid protein VP1 in extracellular structures adjacent to poliovirus-infected cells [44].

Viruses can also manipulate autophagy to dampen key molecules of the innate as well as the inflammatory immune response. Murine cytomegalovirus (MCMV) infection-induced autophagy was shown to selectively target the NF-kB essential modulator (NEMO), via its interaction with the viral protein M45; this leads to the degradation of NEMO in lysosomes and the inhibition of the inflammatory response [45]. UPR induced by HCV activates the complete autophagy pathway, which is required for efficient viral replication and correlates with suppression of innate antiviral immunity [23]. An analogous pathway has been described for DENV infection revealing a novel mechanism to evade an antiviral innate immune response [46].

## 10.4 Conclusions

Organisms have developed highly complex defense networks against invading viruses, and in turn, viruses have also developed mechanisms to disable or manipulate host defenses to hijack their host's defenses. As described in this chapter, autophagy is a clear example of these mechanisms, because depending of the type of viral infection, it can either contribute to limit or promote viral replication. Further studies about the role of autophagy in viral infection will increase our understanding about pathogen–host interactions and open new avenues for the development of novel prophylactic or therapeutic antiviral strategies.

**Acknowledgments** *Funding*: This work was supported by project grants GRUPIN14-071 and GRUPIN14-099 from PCTI (Principado de Asturias) and FEDER funds (EU).

## References

1. Galluzzi L, Pietrocola F, Levine B, Kroemer G. Metabolic control of autophagy. Cell. 2014;159:1263–76.
2. Lamb CA, Yoshimori T, Tooze SA. The autophagosome: origins unknown, biogenesis complex. Nat Rev Mol Cell Biol. 2013;14:759–74.
3. Chiramel AI, Brady NR, Bartenschlager R. Divergent roles of autophagy in virus infection. Cells. 2013;2:83–104.
4. Paulus GL, Xavier RJ. Autophagy and checkpoints for intracellular pathogen defense. Curr Opin Gastroenterol. 2015;31:14–23.
5. Joubert PE, Werneke S, de la Calle C, Guivel-Benhassine F, Giodini A, Peduto L, Levine B, Schwartz O, Lenschow D, Albert ML. Chikungunya-induced cell death is limited by ER and oxidative stress-induced autophagy. Autophagy. 2012;8:1261–3.
6. Joubert PE, Meiffren G, Grégoire IP, Pontini G, Richetta C, Flacher M, Azocar O, Vidalain PO, Vidal M, Lotteau V, Codogno P, Rabourdin-Combe C, Faure M. Autophagy induction by the pathogen receptor CD46. Cell Host Microbe. 2009;6:354–66.
7. Denizot M, Varbanov M, Espert L, Robert-Hebmann V, Sagnier S, Garcia E, Curriu M, Mamoun R, Blanco J, Biard-Piechaczyk M. HIV-1 gp41 fusogenic function triggers autophagy in uninfected cells. Autophagy. 2008;4:998–1008.
8. Espert L, Denizot M, Grimaldi M, Robert-Hebmann V, Gay B, Varbanov M, Codogno P, Biard-Piechaczyk M. Autophagy is involved in T cell death after binding of HIV-1 envelope proteins to CXCR4. J Clin Invest. 2006;116:2161–72.
9. Mikula I Jr, Pastoreková S, Mikula I Sr. Toll-like receptors in immune response to the viral infections. Acta Virol. 2010;54:231–45.
10. Loo YM, Gale M Jr. Immune signaling by RIG-I-like receptors. Immunity. 2011;34:680–92.
11. Fensteri V, Sen GC. Interferons and viral infections. Biofactors. 2009;35:14–20.
12. Sadler AJ, Williams BR. Interferon-inducible antiviral effectors. Nat Rev Immunol. 2008;8:559–68.
13. Gorbea C, Makar KA, Pauschinger M, Pratt G, Bersola JL, Varela J, David RM, Banks L, Huang CH, Li H, Schultheiss HP, Towbin JA, Vallejo JG, Bowles NE. A role for Toll-like receptor 3 variants in host susceptibility to enteroviral myocarditis and dilated cardiomyopathy. J Biol Chem. 2010;285:23208–23.
14. Lee HK, Lund JM, Ramanathan B, Mizushima N, Iwasaki A. Autophagy-dependent viral recognition by plasmacytoid dendritic cells. Science. 2007;315:1398–401.

15. Into T, Inomata M, Takayama E, Takigawa T. Autophagy in regulation of Toll-like receptor signaling. Cell Signal. 2012;24:1150–62.
16. Delgado MA, Deretic V. Toll-like receptors in control of immunological autophagy. Cell Death Differ. 2009;16:976–83.
17. Shi CS, Kehrl JH. MyD88 and Trif target Beclin 1 to trigger autophagy in macrophages. J Biol Chem. 2008;283:33175–82.
18. Lippai M, Lőw P. The role of the selective adaptor p62 and ubiquitin-like proteins in autophagy. Biomed Res Int. 2014;2014:832704.
19. Orvedahl A, MacPherson S, Sumpter R Jr, Talloczy Z, Zou Z, Levine B. Autophagy protects against Sindbis virus infection of the central nervous system. Cell Host Microbe. 2010;7:115–27.
20. B'chir W, Maurin AC, Carraro V, Averous J, Jousse C, Muranishi Y, Parry L, Stepien G, Fafournoux P, Bruhat A. The eIF2α/ATF4 pathway is essential for stress-induced autophagy gene expression. Nucleic Acids Res. 2013;41:7683–99.
21. Martinon F. The endoplasmic reticulum: a sensor of cellular stress that modulates immune responses. Microbes Infect. 2012;14:1293–300.
22. Gade P, Ramachandran G, Maachani UB, Rizzo MA, Okada T, Prywes R, Cross AS, Mori K, Kalvakolanu DV. An IFN-γ-stimulated ATF6-C/EBP-β-signaling pathway critical for the expression of death associated protein kinase 1 and induction of autophagy. Proc Natl Acad Sci U S A. 2012;109:10316–21.
23. Ke PY, Chen SS. Activation of the unfolded protein response and autophagy after hepatitis C virus infection suppresses innate antiviral immunity in vitro. J Clin Invest. 2011;121:37–56.
24. Mohl BP, Tedbury PR, Griffin S, Harris M. Hepatitis C virus-induced autophagy is independent of the unfolded protein response. J Virol. 2012;86:10724–32.
25. McLean JE, Wudzinska A, Datan E, Quaglino D, Zakeri Z. Flavivirus NS4A-induced autophagy protects cells against death and enhances virus replication. J Biol Chem. 2011;286:22147–59.
26. Carpenter JE, Jackson W, Benetti L, Grose C. Autophagosome formation during varicella zoster virus infection following endoplasmic reticulum stress and the unfolded protein response. J Virol. 2011;85:9414–24.
27. Lee J, Giordano S, Zhang J. Autophagy, mitochondria and oxidative stress: cross-talk and redox signalling. Biochem J. 2012;441:523–40.
28. Kosmider B, Messier EM, Janssen WJ, Nahreini P, Wang J, Hartshorn KL, Mason RJ. Nrf2 protects human alveolar epithelial cells against injury induced by influenza A virus. Respir Res. 2012;13:43.
29. Olagnier D, Peri S, Steel C, van Montfoort N, Chiang C, Beljanski V, Slifker M, He Z, Nichols CN, Lin R, Balachandran S, Hiscott J. Cellular oxidative stress response controls the antiviral and apoptotic programs in dengue virus-infected dendritic cells. PLoS Pathog. 2014;10:e1004566.
30. Schaedler S, Krause J, Himmelsbach K, Carvajal-Yepes M, Lieder F, Klingel K, Nassal M, Weiss TS, Werner S, Hildt E. Hepatitis B virus induces expression of antioxidant response element-regulated genes by activation of Nrf2. J Biol Chem. 2010;285:41074–86.
31. Munz C. Antigen processing by macroautophagy for MHC presentation. Front Immunol. 2011;2:42.
32. Leung CS, Taylor GS. Nuclear shelter: the influence of subcellular location on the processing of antigens by macroautophagy. Autophagy. 2010;6:510–1.
33. Paludan C, Schmid D, Landthaler M, Vockerodt M, Kube D, Tuschl T, Münz C. Endogenous MHC class II processing of a viral nuclear antigen after autophagy. Science. 2005;307:593–6.
34. English L, Chemali M, Duron J, Rondeau C, Laplante A, Gingras D, Alexander D, Leib D, Norbury C, Lippé R, Desjardins M. Autophagy enhances the presentation of endogenous viral antigens on MHC class I molecules during HSV-1 infection. Nat Immunol. 2009;10:480–7.
35. Miller S, Krijnse-Locker J. Modification of intracellular membrane structures for virus replication. Nat Rev Microbiol. 2008;6:363–74.

36. Schlegel A, Giddings TH Jr, Ladinsky MS, Kirkegaard K. Cellular origin and ultrastructure of membranes induced during poliovirus infection. J Virol. 1996;70:6576–88.
37. Huang SC, Chang CL, Wang PS, Tsai Y, Liu HS. Enterovirus 71-induced autophagy detected in vitro and in vivo promotes viral replication. J Med Virol. 2009;81:1241–52.
38. O'Donnell V, Pacheco JM, LaRocco M, Burrage T, Jackson W, Rodriguez LL, Borca MV, Baxt B. Foot-and-mouth disease virus utilizes an autophagic pathway during viral replication. Virology. 2011;410:142–50.
39. Granato M, Santarelli R, Farina A, Gonnella R, Lotti LV, Faggioni A, Cirone M. Epstein-barr virus blocks the autophagic flux and appropriates the autophagic machinery to enhance viral replication. J Virol. 2014;88:12715–26.
40. Wen HJ, Yang Z, Zhou Y, Wood C. Enhancement of autophagy during lytic replication by the Kaposi's sarcoma-associated herpesvirus replication and transcription activator. J Virol. 2010;84:7448–58.
41. Grégoire IP, Richetta C, Meyniel-Schicklin L, Borel S, Pradezynski F, Diaz O, Deloire A, Azocar O, Baguet J, Le Breton M, Mangeot PE, Navratil V, Joubert PE, Flacher M, Vidalain PO, André P, Lotteau V, Biard-Piechaczyk M, Rabourdin-Combe C, Faure M. IRGM is a common target of RNA viruses that subvert the autophagy network. PLoS Pathog. 2011;7:e1002422.
42. Wang X, Hou L, Du J, Zhou L, Ge X, Guo X, Yang H. Capsid, membrane and NS3 are the major viral proteins involved in autophagy induced by Japanese encephalitis virus. Vet Microbiol. 2015;178:217–29.
43. Li J, Liu Y, Wang Z, Liu K, Wang Y, Liu J, Ding H, Yuan Z. Subversion of cellular autophagy machinery by hepatitis B virus for viral envelopment. J Virol. 2011;85:6319–33.
44. Jackson WT, Giddings TH Jr, Taylor MP, Mulinyawe S, Rabinovitch M, Kopito RR, Kirkegaard K. Subversion of cellular autophagosomal machinery by RNA viruses. PLoS Biol. 2005;3:e156.
45. Fliss PM, Jowers TP, Brinkmann MM, Holstermann B, Mack C, Dickinson P, Hohenberg H, Ghazal P, Brune W. Viral mediated redirection of NEMO/IKKgamma to autophagosomes curtails the inflammatory cascade. PLoS Pathog. 2012;8:e1002517.
46. Green AM, Beatty PR, Hadjilaou A, Harris E. Innate immunity to dengue virus infection and subversion of antiviral responses. J Mol Biol. 2014;426:1148–60.

# Chapter 11
# Cancer Stem Cells and Autophagy: Present Knowledge and Future Perspectives

**Bakiye Goker Bagca and Cigir Biray Avci**

**Abstract** Macroautophagy, commonly referred to as autophagy, is a recycling process involving lysosomal degradation of the cell components such as proteins and organelles. This process prevents damage to the cell through the degradation of nonfunctional cellular components and provides raw material and energy which are required to realize biosynthesis reactions. Since autophagy has evolutionarily conserved complex molecular mechanisms, the relationship between autophagy and cancer is multifaceted. There are some insights in which autophagy, also referred to as type 2 cell death, has been suggested as an alternative approach to kill cancer cells have defected apoptosis mechanism. On the other hand, it has also been shown in recent studies that autophagy mechanism, especially in cancer stem cells, may be responsible for obtaining epithelial–mesenchymal transition, invasion, metastasis, drug resistance and recurrence. This chapter focuses on the role of autophagy mechanisms on cancer stem cells and its place in future treatment approaches.

## 11.1 Introduction

The odyssey of autophagy which led Yoshinori Ohsumi to be awarded with Nobel Prize in Physiology or Medicine in 2016 began with Christian de Duve's discovery of lysosomes in 1955 [1, 2]. The term autophagy, which consists of the words "self" (auto) and "eating" (phagy), was also coined by Christian de Duve in 1966 [3]. According to their function and distribution of their cargo to lysosomes, three different autophagy types are defined as "*macroautophagy*," "*microautophagy*," and "*chaperone-mediated autophagy*" [4]. Microautophagy is characterized as the recycling process of intracellular components, usually long-lived proteins, through non-specific lysosomal degradation [5]. Proteins which are transported to the lysosomal membrane through the heat shock protein family chaperones and they are digested in chaperone-mediated autophagy to prevent proteotoxicity [6]. However,

Authors Bakiye Goker Bagca and Cigir Biray Avci equally contributed to this work.

B. Goker Bagca · C. Biray Avci (✉)
Department of Medical Biology, Faculty of Medicine, Ege University, Izmir, Turkey



K. Turksen (ed.), *Autophagy in Health and Disease*, Stem Cell Biology and Regenerative Medicine, https://doi.org/10.1007/978-3-319-98146-8_11

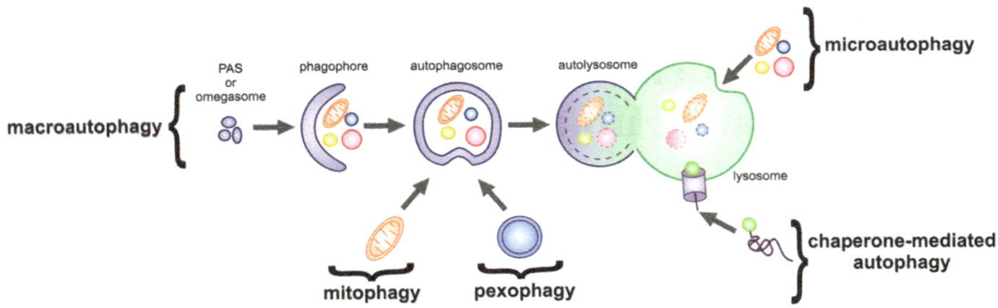


**Fig. 11.1** Types of autophagy mechanisms

autophagy term is usually used to describe macroautophagy which is characterized by the formation of "autophagosomes" for lysosomal degradation of cytoplasmic macromolecules such as organelles [7]. In addition to these subtypes, specifically degradation of mitochondria and peroxisome by autophagic mechanisms are defined as "*mitophagy*" and "*pexophagy*," respectively [Fig. 11.1] [8, 9].

Autophagy was firstly described by its morphological characteristics. The phagophore originating from the endoplasmic reticulum associated structure named "omegasome" or "phagophore assembly site (PAS)" expands around the material to be degraded and forms a characteristic double-membrane vesicle called the "autophagosome." Mature autophagosome forms "autolysosome" via fusion with lysosomes. The material in the autolysosomes is degraded by lysosomal hydrolases [Fig. 11.1] [10]. The molecular mechanism of autophagy has begun to be illuminated by the discovery of the *ATG*, initially called as *APG*, genes which encode the "*autophagy-related proteins*" that are responsible for de novo autophagosome formation in yeast. Since *ATG* proteins organize multiprotein complexes and participate in all stages of autophagy, inhibition of even just one *ATG* gene prevents the autophagic flux. The most important key of autophagy is the *mTORC1* (Mechanistic Target of Rapamycin) serine–threonine kinase protein complex, which responds to nutrient levels. In the physiological state, this complex suppresses the *ULK1* (Unc-51 Like Autophagy Activating Kinase 1) protein complex that initiates autophagy. Autophagy may be initiated by the *AMPK* (AMP-activated protein kinase) which activates *ULK1* protein complex via directly or inhibiting *MTORC1* complex. This complex initiates nucleation step. The BECN-VPS complex, including also *ATG14* and *ATG9* proteins, triggers phagophore formation. Phagophore formation may be inhibited by the *BCL-2* (B-Cell CLL/Lymphoma) family. The complex consisting of *ATG12*, *ATG5*, and *ATG16* allows both phagophore elongation and translocation of LC3-II, which is formed via *ATG4, ATG7*, and *ATG3*, to the autophagosome membrane [Fig. 11.2] [11–13].

Autophagy, like all other genetic mechanisms, is regulated epigenetically. The most important epigenetic regulators are noncoding RNAs (ncRNA) which include microRNAs (miRNAs) are about 20 nucleotides and long-noncoding RNAs (lncRNAs) are longer than 200 nucleotides that regulate gene expression. The ncRNAs usually control autophagy by regulating expression level of ATG genes.

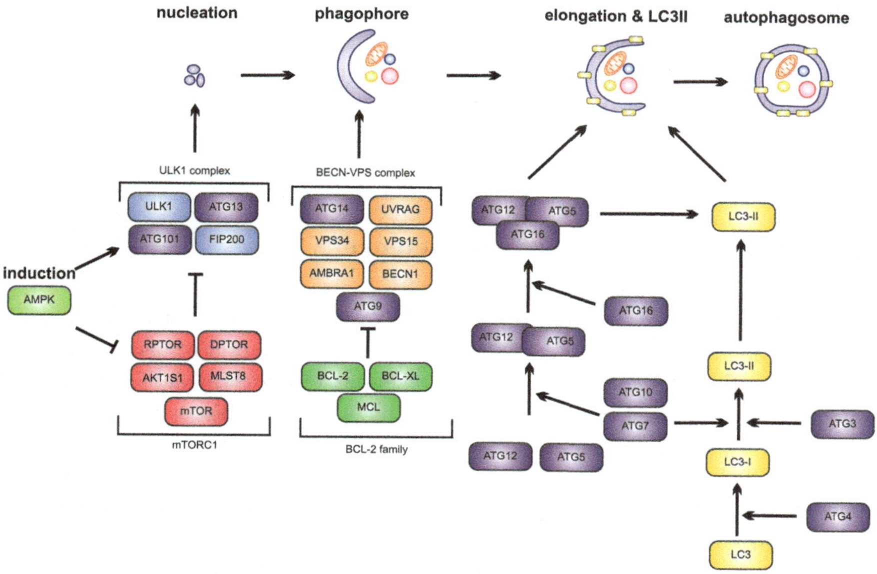


**Fig. 11.2** Molecular mechanisms of autophagy (macroautophagy)

The most known autophagy regulator miRNAs are listed: *miR-100, miR-101, miR-106, miR-10a, miR-124, miR-125b1, miR-126, miR-129, miR-130a, miR-137, miR-138, miR-140-5p, miR-143, miR-144, miR-152, miR-155, miR-155-3p, miR-16, miR-17-5p, miR-181a, miR-183, miR-18a, miR-193b, miR-199A-5p, miR-19b, miR-200b, miR-200c, miR-204, miR-205, miR-20a, miR-21, miR-212, miR-214, miR-216a, miR-216b, miR-218, miR-22, miR-224, miR-224-3p, miR-23a, miR-23B-3p, miR-24-3p, miR-25, miR-26a, miR-290-295, miR-29a, miR-29b, miR-30a, miR-30d, miR-32, miR-340, miR-34a, miR-372, miR-373, miR-374a, miR-375, miR-376a, miR-376b, miR-409-3p, miR-4487, miR-451, miR-451a, miR-487b-5p, miR-502, miR-5195-3p, miR-519a, miR-595, miR-630, miR-634, miR-638, miR-7, miR-885-3p, miR-93, miR-9-3p, miR-96, miR-let7f1* [14]. The current known autophagic regulator lncRNAs are also *NBR2, Ad5-AlncRNA, loc146880, Linc-ROR, MEG3, TGFB2-OT1/FLJ11812, HULC, PCGEM1, AK156230, PTENP1, Chast, MALAT, HOTAIR, GAS5, HNF1A-AS1* [15].

Autophagy is an evolutionarily conserved, two-faced, complex process that must be tightly regulated by cellular signaling pathways. Autophagy functions differently from apoptosis, both prosurvival and prodeath mechanisms. In stress situations such as starvation or hypoxia, the lysosomal degradation of cellular materials provides the energy and raw material needed to keep the cell alive. However, if autophagy is unable to keep the cell alive, it triggers "programmed-type II cell death" [16].

Since autophagy plays a fundamental role in the homeostatic process, it is closely related to many pathological situation as well as the physiological conditions. Abnormal autophagy flux has been associated with numerous pathologies including cancer, aging, metabolic, cardiovascular, pulmonary, infectious, and neurodegenerative disorders. Because of the illuminating of the molecular mechanism of autophagy is critical for treatment of these diseases, many studies have been conducted to clarify the autophagic flux. Selectively targeting of autophagy induction or inhibition is in the focus of molecular and clinical researches, particularly to overcome cancer stem cells [17].

## 11.2 Stem Cell: Queen of Competence

To achieve homeostasis in multicellular organisms, the death of cells have potency to damage the organism must be balanced by the proliferation of new cells. Stem cells are undifferentiated cells which have potential to transform and generate to all cell types of specialized tissues [18]. Stem cells were firstly discovered in 1961 by Till and McCulloch as progenitor cells in the hematopoietic system [19].

Stem cells have the potential to produce different tissue types with their self-renewal, pluripotency, clonality, differentiation, plasticity, transdifferentiation, and stemness properties. Self-renewal means the proliferation of the stem cell without any differentiation, thereby the stem cell functions as a stem cell source which can continuously produce new stem cells. Pluripotency term defines the capacity of the stem cell to differentiate into which cells it produces [20]. Stem cells have totipotent, pluripotent, multipotent, or unipotent properties according to their potency [21]. Zygote, two- and four-cell stage blastomeres are totipotent stem cells. Each totipotent stem cell has the potential to form a complete organism by itself [22]. Differentiation capacity of stem cells is gradually being limited during developmental process according to extrinsic signals. Embryonic stem cells are pluripotent which have differentiation potential to all cell types in the organism. However, embryonic stem cells are not able to generate a whole organism alone [23]. Multipotent stem cells are undifferentiated cells in the adult tissue and they have limited differentiation capacities which generate specific cell types of a certain tissue. Mesenchymal stem cells, hematopoietic stem cells, and endothelial progenitor cells in adult tissues are multipotent stem cells and they produce cells when needed [24]. The stem cells that can differentiate into one type of cell are unipotent [25]. In addition, induced pluripotent stem cells (iPSC) were also produced by induction of fibroblasts using pluripotency factors [26].

The differentiation characteristics of stem cells are controlled by genetic and epigenetic mechanisms [27, 28]. *OCT4* (POU homeodomain transcription factor), *SOX*2 (SRY-related HMG-box transcription factor), and *NANOG* (DNA binding homeobox transcription factor) transcription factors were defined as essential pluripotency factors in embryonic stem cells [29–31]. Positive feedback loop between the pluripotency factors is major regulator of self-renewal. *LIF* (interleukin 6 family cytokine)/*JAK* (Janus Kinase)/*STAT3* (signal transducer and activator of transcription 3), *BMP* (bone morphogenetic protein)/*SMAD* (Sma- and Mad-related protein)/*ID* (inhibitor of dna binding), *FGF* (fibroblast growth factor)/*PI3K* (phosphatidylinositol-4,5-bisphosphate 3-kinase)/*AKT*1 (AKT serine/threonine kinase1) signaling pathways and *FOXO* (forkhead box O), *TP53* (tumor protein P53) proteins also play critical role for maintaining of pluripotency through inhibition of *MAPK* (mitogen-activated protein kinase) pathway which induces differentiation [32]. In addition to these oncogenes *MYC* (proto-oncogene, BHLH transcription factor) and *KLF4* (Kruppel like factor 4) are also pluripotency factors which are used for inducing iPSC by Yamanaka [26].

Moreover, the strict regulation of the cell cycle is necessary to the limited differentiation ability and "quiescent" characteristic in adult stem cells. Cell cycle regulators such as cyclins, cyclin-dependent kinases, and cyclin-dependent kinase inhibitors are critical for regulation of this differentiation process [33]. *WNT* (Wingless-Type MMTV Integration Site Family), Hedgehog, Notch signaling pathways are responsible for differentiation of progenitor cells appropriately, as well as self-renewal of stem cells, during embryogenesis. These mitogenic signaling pathways which compose of proto-oncogenes are tightly controlled in adult tissues by tumor-suppressors [34–36].

## 11.3 Cancer Stem Cell: More Than a Stem Cell

Cancer stem cells originate from the genetic and epigenetic abnormalities of normal stem cells or differentiated cells in normal tissues [37, 38]. Cancer stem cells are analogs of normal stem cells which responsible for the production and differentiation of new cells to maintain tissue homeostasis. Because it is responsible for the onset of the tumor, it is also called "tumor-initiating cells." However, it is more commonly called "cancer stem cell" in order to emphasize its role in tumor heterogeneity [Fig. 11.3] [39]. Chronologically, it was first noticed that heterogeneity in the tumor bulk, and then a small number of cells in the tumor mass had the potential to initiate tumors independently. Subsequently, when heterogeneous tumor tissue was examined, it was discovered that cells were present in undifferentiated state. Finally, CSCs were described the cells only have $CD34^{+}CD38^{-}$ surface marker pattern may initiate new tumor by Bennet and Dick in acute myeloid leukemia [40].

CSCs have all the properties of normal stem cells such as self-renewal, clonality, and stemness. In addition to these abilities, CSCs have the ability to slow proliferation which plays a role in treatment resistance, dormancy (quiescence), and relapse [41]. CSCs acquire these distinctiveness through the abnormal activation of the Notch, Hedgehog, and WNT signaling pathways which are the developmental regulators of embryonic stem cells [42–44]. In addition, aberrations of the JAK/STAT, PI3K, *NFKB* (nuclear factor kappa B subunit) signaling pathways, involved in normal tissue homeostasis and their cross talks are also important in CSC maintaining [45]. Epithelial–mesenchymal transition (EMT) and the escape of anoikis also support the cancer cells in metastasis and chemotherapy/radiotherapy resistance as well as tumor initiation [46]. EMT is a developmental, reversible process that allows epithelial cells to transform into mesenchymal cells during embryogenesis. Therefore, cells that are able to motility can migrate to distant regions and generate different tissues and organs. This process is particularly tightly controlled by E-cadherin, β-catenin, and

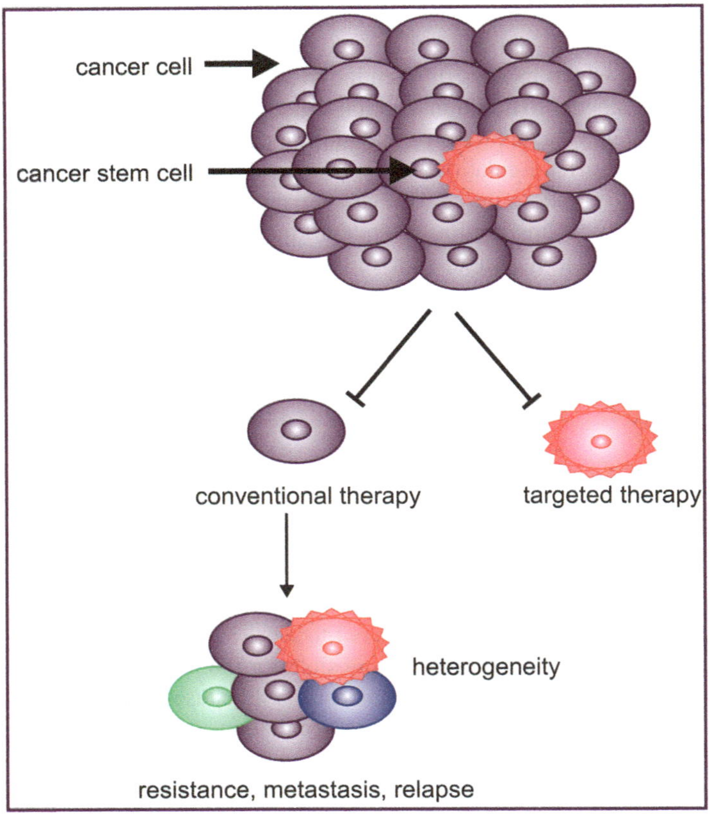


**Fig. 11.3** Cancer stem cell

especially *TGFβ* (Transforming Growth Factor Beta) signaling [47]. Anoikis is a special form of cell death in which cells that separate from their extracellular matrix undergo apoptosis through cell adhesion molecules such as integrins and cadherins. Thus, the cells are prevented from being placed in different tissues [48]. It has been suggested that EMT induction, particularly due to anoikis resistance, plays a critical role in cancer stem cell motility and secondary tumor focus formation [49].

## 11.4 Autophagy in Cancer: Same Pathway, Different Effect

The dual role of autophagy which makes a choice between the qualities of either promoting or suppressing cancer is determined according to the type and grade/stage of the cancer that is dependent on physiological conditions, like normal cells [50]. Initially, it has been determined that disregulations and loss-of-function/heterozygosity mutations of the *BECN1* (Beclin 1), *UVRAG* (UV radiation resistance associated), *PTEN* (phosphatase and tensin homolog), *AMPK*, *LKB1* (serine/threonine kinase 11), and *TSC1/2* (tuberous sclerosis) genes which are positive regulators of autophagy, play role in the initiation and promotion of cancer. These genes are referred to as tumor suppressors or candidates. Moreover, oncogenes such as *RAS* (proto-oncogene, GTPase), *RHEB* (Ras homolog, MTORC1 binding), and *AKT1* are defined as negative regulators of autophagy [51]. For these reasons, autophagy is mainly regarded as a tumor suppressor mechanism. On the other hand, there are numerous studies showing that autophagy is also related to uncontrolled cell proliferation, invasion, angiogenesis, metastasis, and escape from programmed cell death in different cancer types. It is crucial to understand the autophagy mechanism in CSCs since these cells are the primary actors in the tumor initiation, progression, metastasis, and treatment resistance.

## 11.5 Autophagy and CSC: Paradoxical Couple

It is known that autophagic flux promotes the stemness feature which allows the stem cells maintaining through eliminating potential harmful substances in autolyosomes. However, this flux is necessary to maintain the continuity of CSCs as well as "normal" stem cells [52].

Autophagy affects the energy dynamics of the CSCs which leads to the development of adaptive behavior of cancer stem cells by altering the tumor microenvironment. Especially hypoxia induced autophagy has been suggested to provide advantages to CSCs in nutrition recycling as well as tumor microenvironment alteration which affects the cellular signaling pathways and promotes to CSC continuity and evolution [53].

The relationship between autophagy and the role of different CSCs in tumor initiation, progression, and aggressiveness is emphasized by evidences obtained from scientific studies.

### *11.5.1 Breast Cancer*

Autophagy has been associated with stem cell formation, continuity, metastasis, and chemotherapy resistance in different types of breast cancer. It has been shown that autophagy induction through *ATG4A* and *BECN1* genes promotes stem cell formation and tumorigenicity of CSCs and as a result of this autophagy has been suggested to play an important role in chemotherapy resistance in *HER2* (Erb-B2 receptor tyrosine kinase) overexpressing breast cancer [54–56]. It is also reported that autophagy promotes breast cancer stem cell continuation through interleukin levels especially *IL6* (Interleukin 6) [57]. It has been reported that autophagy causes dormancy by regulating *MAPK8* signaling in breast cancer stem cells [58]. In *HER2* overexpressing breast cancer, as well as in triple negative breast cancer where there is lack of expression of *HER2*, estrogen, and progesterone receptors, autophagy has been shown to play an essential role in stem cell development leading to chemotherapy [59]. From this point of view, it maybe suggested that autophagy is associated with breast cancer stem cells independent of tumor status. Moreover, the negative effects of inhibition of autophagy on breast CSCs have been illuminated. Suppression of epithelial-to-mesenchymal transition through autophagy inhibition reduces stem cell frequency in breast cancer [60]. Autophagic flux inhibits the effects of cytotoxic agents on *HER2* overexpressing breast cancer cells and it provokes to occurence of chemotherapy resistant cells [61]. On the other hand, the conflicting state of autophagy is also evident in breast CSCs. In a study involving hormone-independent mammary cancer cells, it has been shown that regulating autophagy through *PKCD* (Protein Kinase C Delta) has dual role which includes both reduction of cancer stem cell population and induction of self-renewal property [62]. Moreover, it has been also suggested that lower autophagy level may be promote stemness and lead to poor prognosis in triple-negative breast cancer stem cells [63]. This situation can be explained that autophagy is suggested as a fundamental cell death mechanism in some situations for cancer stem cells which have insufficient apoptotic cell death [64].

### *11.5.2 Glioblastoma*

*DRAM1* (DNA damage regulated autophagy modulator) and *SQSTM1* (sequestosome 1), basic regulators of autophagy were defined as crucial players in glioblastoma stem cells [65]. In addition to this, the silencing of the *ATG7* gene of autophagy in glioblastoma cancer stem cells supports the efficacy of chemotherapy [66].

### *11.5.3 Pancreatic Cancer*

Autophagy induction has been shown to play an essential role in the development of pancreatic cancer. It has been shown that primer pancreatic cells and pancreatic cell lines have increased level of autophagy which is essential for tumour initiation and development [67]. This situation is particularly associated with cancer stem cells. Autophagy suppression via knockdown of *ATG5* and *ATG7* induces the efficiency of antiproliferative agent used in the CD133 positive stem cell fraction in pancreatic cancer [68]. Autophagy inhibition has been shown to decrease and regress in the pancreatic cancer stem cell population [69].

### 11.5.4 Gynecologic Cancers

Autophagy induction promotes the stemness and self-renewal properties of stem cells through the upregulation of *FOXA2* (forkhead box A2) gene in ovarian cancer stem cells [70]. In the cervical cancer mouse model, the autophagic pathway promotes stem cell stemness and tumor progression via expression of beclin-1 which controls the protein level of pluripotency-associated transcription factors such as *OCT4*, *SOX2*, and *NANOG* [71]. Autophagy flux is defined as one of the critical process which supports chemoresistance and stem cell stemness in endometrial cancers [72].

### 11.5.5 Bladder Cancers

It is reported that the inhibition of autophagy inhibits stem cell progression in bladder cancer [73]. In a study conducted with urothelial carcinoma stem cells, autophagy inhibition has been shown to reduce the expression levels of glycolysis genes, drug resistance genes (*ABCG2*, ABCB1 (ATP binding cassette subfamily)), and stemness genes (*OCT4, NANOG*). Therefore, autophagy inhibition has been defined as an important target to overcome drug resistance in urothelial carcinoma [74].

### 11.5.6 Hepatocellular Carcinoma/Liver Cancer

In a mouse model with liver cirrhosis, abnormal autophagy causes the hepatocellular carcinoma via stimulating the transition of normal hepatic cells into CD90+ liver CSCs. It is suggested that the activation of the *MET* (proto-oncogene, receptor tyrosine kinase)/*MAPK8* and *MET/STAT3* pathways through *HGF* (hepatocyte growth factor) overexpression following autophagy induction plays a key role in this process [75]. Conversely, liver CSCs have also been shown to be resistant to interferon-gamma-induced autophagy [76].

### 11.5.7 Colorectal Cancer

Colorectal cancer is one of the most malignant tumors with a highly metastatic character which originate from cancer stem cells. The induction of autophagy in colorectal CSC contributes to tumor development through *PROX1* (Prospero Homeobox 1) [77]. Epigenetic regulators also play a role in colorectal cancer. Overexpressed hsa-miR-140-5p which targets a critical autophagosome formation gene *ATG12* and *SMAD2* (SMAD Family Member2) gene reduces cell viability by inhibiting autophagy in colorectal cancer stem cells [78].

### *11.5.8 Renal Cell Carcinoma*

In renal cell carcinoma cells, induction of autophagy causes formation of stem cell-like phenotype through stimulating EMT [79]. Autophagy is also defined as a biomarker and it is suggested as a potential therapy target in renal cell carcinoma CSCs [80].

### *11.5.9 Lung Cancer*

Inhibition of autophagy enhances the cytotoxic and apoptotic effects of chemotherapeutic agent in stem-like lung cancer cells [81].

### *11.5.10 Neuroblastoma*

It is shown that autophagic degradation of Id proteins (inhibitor of differentiation 1/2) and *BECN1* were triggered by *CaMKII* (Calcium/calmodulin-dependent protein kinase II) regulates neuroblastoma stem cell number via regulating abnormal cell differentiation in neuroblastoma [82].

## 11.6 Conclusions and Future Perspective

The preclinical use of autophagy inhibitors such as 3-methyladenine, wortmannin, chloroquine and hydroxychloroquine alone or in combination with chemotherapy/radiotherapy for eliminating of various cancer cells has been suggested to increase the efficacy of the treatment [83]. It has been also reported that the anticancer agents treated to breast, colorectal, hepatocellular carcinoma, urothelial, renal, cervical, glioma, glioblastoma, ovarian, pancreatic, prostate, liver, lung, leukemia, myeloma, oral cancer and rhabdomyosarcoma cells induce autophagic death [84, 85]. However, these studies only involve cancer cell lines that have a homogenous genetic structure and do not contain CSCs that cause cancer heterogeneity. CSCs, called tumor-initiating cells, are originated from genetic or epigenetic alterations of normal stem cells, normal cells or cancer cells. CSCs are also responsible for tumor heterogeneity, progression, invasion, metastasis, chemotherapy and radiotherapy resistance and relapse, as well as the tumor initiation. The abnormal activation of signaling pathways and metabolic rearrangements have been occurred in CSCs allow cancer to gain tumorigenic advantages. Increased glycolysis in cancer cells is an adaptation mechanism to meet the energy need. Similarly, the enhanced autophagy may also provide both nutrients and energy support to the cancer cells. Besides nutritional support, it has been shown that autophagy also gives CSCs advantage over hypoxia and metabolic stress situations like chemotherapy or radiotherapy. For this reason, inhibition of autophagy has begun to be regarded as a new therapeutic approach with the elimination of CSCs which are the most critical target in cancer treatment. Autophagy is also anticipated as a tumor suppressor mechanism since it is involved in the degradation of potentially destructive materials for the normal cell homeostasis. Moreover, autophagy, named type II cell death, is an alternative targeted therapy approach for eliminating CSCs which have disregulated apoptosis mechanism.

For these reasons, it is still contradictory whether it is the induction or inhibition of autophagy that should be aimed at in cancer treatment. Additional work is needed to decide whether to inhibit or activate autophagy to explore its potential role in cancer therapy.

## References

1. The Nobel Prize in Physiology or Medicine 2016. Nobelprize.org. Nobel Media AB 2014. Web. 16 Dec 2017. http://www.nobelprize.org/nobel_prizes/medicine/laureates/2016/.
2. De Duve C, Pressman BC, Gianetto R, Wattiaux R, Appelmans F. Tissue fractionation studies. 6. Intracellular distribution patterns of enzymes in rat-liver tissue. Biochem J. 1955;60(4):604–17.
3. de Duve C, Wattiaux R. Functions of lysosomes. Annu Rev Physiol. 1966;28(1):435–92.
4. Mizushima N, Komatsu M. Autophagy: renovation of cells and tissues. Cell. 2011;147(4):728–41.
5. Mijaljica D, Prescott M, Devenish RJ. Microautophagy in mammalian cells: revisiting a 40-year-old conundrum. Autophagy. 2011;7(7):673–82.
6. Catarino S, Pereira P, Girão H. Molecular control of chaperone-mediated autophagy. Essays Biochem. 2017;61(6):663–74.
7. Mizushima N, Yoshimori T, Ohsumi Y. The role of Atg proteins in autophagosome formation. Annu Rev Cell Dev Biol. 2011;27:107–32.
8. Lemasters JJ. Selective mitochondrial autophagy, or mitophagy, as a targeted defense against oxidative stress, mitochondrial dysfunction, and aging. Rejuvenation Res. 2005;8(1):3–5.
9. Wang CW, Kim J, Huang WP, Abeliovich H, Stromhaug PE, Dunn WA Jr, Klionsky DJ. Apg2 is a novel protein required for the cytoplasm to vacuole targeting, autophagy, and pexophagy pathways. J Biol Chem. 2001;276(32):30442–51.
10. Parzych KR, Klionsky DJ. An overview of autophagy: morphology, mechanism, and regulation. Antioxid Redox Signal. 2014;20(3):460–73.
11. Suzuki K, Kirisako T, Kamada Y, Mizushima N, Noda T, Ohsumi Y. The pre-autophagosomal structure organized by concerted functions of APG genes is essential for autophagosome formation. EMBO J. 2001;20(21):5971–81.
12. Suzuki K, Noda T, Ohsumi Y. Interrelationships among Atg proteins during autophagy in Saccharomyces cerevisiae. Yeast. 2004;21(12):1057–65.
13. Mariño G, Niso-Santano M, Baehrecke EH, Kroemer G. Self-consumption: the interplay of autophagy and apoptosis. Nat Rev Mol Cell Biol. 2014;15(2):81–94.

14. Gozuacik D, Akkoc Y, Ozturk DG, Kocak M. Autophagy-regulating microRNAs and cancer. Front Oncol. 2017;7:65.
15. Yang L, Wang H, Shen Q, Feng L, Jin H. Long non-coding RNAs involved in autophagy regulation. Cell Death Dis. 2017;8(10):e3073.
16. Fuchs Y, Steller H. Live to die another way: modes of programmed cell death and the signals emanating from dying cells. Nat Rev Mol Cell Biol. 2015;16(6):329–44.
17. Choi AM, Ryter SW, Levine B. Autophagy in human health and disease. N Engl J Med. 2013;368(7):651–62.
18. Goodell MA, Nguyen H, Shroyer N. Somatic stem cell heterogeneity: diversity in the blood, skin and intestinal stem cell compartments. Nat Rev Mol Cell Biol. 2015;16(5):299–309.
19. Till JE, Mcculloch EA. A direct measurement of the radiation sensitivity of normal mouse bone marrow cells. Radiat Res. 1961;14:213–22.
20. Weissman IL. Stem cells: units of development, units of regeneration, and units in evolution. Cell. 2000;100(1):157–68.
21. Singh VK, Saini A, Kalsan M, Kumar N, Chandra R. Describing the stem cell potency: the various methods of functional assessment and in silico diagnostics. Front Cell Dev Biol. 2016;4:134.
22. Kelly SJ. Studies of the developmental potential of 4- and 8-cell stage mouse blastomeres. J Exp Zool. 1974;200(3):365–76.
23. Yu J, Thomson JA. Pluripotent stem cell lines. Genes Dev. 2008;22(15):1987–97.
24. Liao SY, Tse HF. Multipotent (adult) and pluripotent stem cells for heart regeneration: what are the pros and cons? Stem Cell Res Ther. 2013;4(6):151.
25. Visvader JE, Clevers H. Tissue-specific designs of stem cell hierarchies. Nat Cell Biol. 2016;18(4):349–55.
26. Takahashi K. Yamanaka S (2006) Induction of pluripotent stem cells from mouse embryonic and adult fibroblast cultures by defined factors. Cell. 2006;126(4):663–76.
27. Cheung TH, Rando TA. Molecular regulation of stem cell quiescence. Nat Rev Mol Cell Biol. 2013;14(6):329–40.
28. Avgustinova A, Benitah SA. Epigenetic control of adult stem cell function. Nat Rev Mol Cell Biol. 2013;17(10):643–58.
29. Nichols J, Zevnik B, Anastassiadis K, Niwa H, Klewe-Nebenius D, Chambers I, Schöler H, Smith A. Formation of pluripotent stem cells in the mammalian embryo depends on the POU transcription factor Oct4. Cell. 1998;95(3):379–91.
30. Nishimoto M, Fukushima A, Okuda A, Muramatsu M. The gene for the embryonic stem cell coactivator. UTF1 carries a regulatory element which selectively interacts with a complex composed of Oct-3/4 and Sox-2. Mol Cell Biol. 1999;19(8):5453–65.
31. Chambers I, Colby D, Robertson M, Nichols J, Lee S, Tweedie S, Smith A. Functional expression cloning of Nanog, a pluripotency sustaining factor in embryonic stem cells. Cell. 2003;113(5):643–55.
32. He S, Nakada D, Morrison SJ. Mechanisms of stem cell self-renewal. Annu Rev Cell Dev Biol. 2009;25:377–406.
33. Zhang H, Wang ZZ. Mechanisms that mediate stem cell self-renewal and differentiation. J Cell Biochem. 2008;103(3):709–18.
34. Holland JD, Klaus A, Garratt AN, Birchmeier W. Wnt signaling in stem and cancer stem cells. Curr Opin Cell Biol. 2013;25(2):254–64.
35. Petrova R, Joyner AL. Roles for Hedgehog signaling in adult organ homeostasis and repair. Development. 2014;141(18):3445–57.
36. Liu J, Sato C, Cerletti M, Wagers A. Notch signaling in the regulation of stem cell self-renewal and differentiation. Curr Top Dev Biol. 2010;92:367–409.
37. Bjerkvig R, Tysnes BB, Aboody KS, Najbauer J, Terzis AJ. Opinion: the origin of the cancer stem cell: current controversies and new insights. Nat Rev Cancer. 2005;5(11):899–904.
38. Feinberg AP, Ohlsson R, Henikoff S. The epigenetic progenitor origin of human cancer. Nat Rev Genet. 2006;7(1):21–33.

39. Vermeulen L, Sprick MR, Kemper K, Stassi G, Medema JP. Cancer stem cells--old concepts, new insights. Cell Death Differ. 2008;15(6):947–58.
40. Clevers H. The cancer stem cell: premises, promises and challenges. Nat Med. 2011;17(3):313–9.
41. Kleffel S, Schatton T. Tumor dormancy and cancer stem cells: two sides of the same coin? Adv Exp Med Biol. 2013;734:145–79.
42. Wang J, Sullenger BA, Rich JN. Notch signaling in cancer stem cells. Adv Exp Med Biol. 2012;727:174–85.
43. Cochrane CR, Szczepny A, Watkins DN, Cain JE. Hedgehog signaling in the maintenance of cancer stem cells. Cancers (Basel). 2015;7(3):1554–85.
44. de Sousa E Melo F, Vermeulen L. Wnt signaling in cancer stem cell biology. Cancers (Basel). 2016;8(7):pii: E60.
45. Matsui WH. Cancer stem cell signaling pathways. Medicine (Baltimore). 2016;95(1 Suppl 1):S8–S19.
46. Kruyt FA, Schuringa JJ. Apoptosis and cancer stem cells: implications for apoptosis targeted therapy. Biochem Pharmacol. 2010;80(4):423–30.
47. Puisieux A, Brabletz T, Caramel J. Oncogenic roles of EMT-inducing transcription factors. Nat Cell Biol. 2014;16(6):488–94.
48. Yu LG. Cancer cell resistance to anoikis: MUC1 glycosylation comes to play. Cell Death Dis. 2017;8(7):e2962.
49. Frisch SM, Schaller M, Cieply B. Mechanisms that link the oncogenic epithelial-mesenchymal transition to suppression of anoikis. J Cell Sci. 2013;26(Pt 1):21–9.
50. Singh SS, Vats S, Chia AY, Tan TZ, Deng S, Ong MS, Arfuso F, Yap CT, Goh BC, Sethi G, Huang RY, Shen HM, Manjithaya R, Kumar AP. Dual role of autophagy in hallmarks of cancer. Oncogene. 2018;37:1142.
51. Kung CP, Budina A, Balaburski G, Bergenstock MK, Murphy M. Autophagy in tumor suppression and cancer therapy. Crit Rev Eukaryot Gene Expr. 2011;21(1):71–100.
52. Pan H, Cai N, Li M, Liu GH, Izpisua Belmonte JC. Autophagic control of cell 'stemness'. EMBO Mol Med. 2013;5(3):327–31.
53. Naik PP, Panda PK, Bhutia SK. Oral cancer stem cells microenvironment. Adv Exp Med Biol. 2017;1041:207–33.
54. Wolf J, Dewi DL, Fredebohm J, Müller-Decker K, Flechtenmacher C, Hoheisel JD, Boettcher M. A mammosphere formation RNAi screen reveals that ATG4A promotes a breast cancer stem-like phenotype. Breast Cancer Res. 2013;15(6):R109.
55. Gong C, Bauvy C, Tonelli G, Yue W, Deloménie C, Nicolas V, Zhu Y, Domergue V, Marin-Esteban V, Tharinger H, Delbos L, Gary-Gouy H, Morel AP, Ghavami S, Song E, Codogno P, Mehrpour M. Beclin 1 and autophagy are required for the tumorigenicity of breast cancer stem-like/progenitor cells. Oncogene. 2013;32(18):2261–72. 2272e.1-11.
56. Rodríguez CE, Berardi DE, Abrigo M, Todaro LB, Bal de Kier Joffé ED, Fiszman GL. Breast cancer stem cells are involved in Trastuzumab resistance through the HER2 modulation in 3D culture. J Cell Biochem. 2018;119(2):1381–91.
57. Maycotte P, Jones KL, Goodall ML, Thorburn J, Thorburn A. Autophagy supports breast cancer stem cell maintenance by regulating IL6 secretion. Mol Cancer Res. 2015;13(4):651–8.
58. Chaterjee M, van Golen KL. Breast cancer stem cells survive periods of farnesyl-transferase inhibitor-induced dormancy by undergoing autophagy. Bone Marrow Res. 2011;2011:362938.
59. Bousquet G, El Bouchtaoui M, Sophie T, Leboeuf C, de Bazelaire C, Ratajczak P, Giacchetti S, de Roquancourt A, Bertheau P, Verneuil L, Feugeas JP, Espié M, Janin A. Targeting autophagic cancer stem-cells to reverse chemoresistance in human triple negative breast cancer. Oncotarget. 2017;8(21):35205–21.
60. Zhang L, Xu L, Zhang F, Vlashi E. Doxycycline inhibits the cancer stem cell phenotype and epithelial-to-mesenchymal transition in breast cancer. Cell Cycle. 2017;16(8):737–45.
61. Rodríguez CE, Reidel SI, Bal de Kier Joffé ED, Jasnis MA, Fiszman GL. Autophagy protects from trastuzumab-induced cytotoxicity in HER2 overexpressing breast tumor spheroids. PLoS One. 2015;10(9):e0137920.

62. Berardi DE, Flumian C, Rodriguez CE, Bessone MI, Cirigliano SM, Joffé ED, Fiszman GL, Urtreger AJ, Todaro LB. PKCδ inhibition impairs mammary cancer proliferative capacity but selects cancer stem cells, involving autophagy. J Cell Biochem. 2016;117(3):730–40.
63. Chang SJ, Ou-Yang F, Tu HP, Lin CH, Huang SH, Kostoro J, Hou MF, Chai CY, Kwan AL. Decreased expression of autophagy protein LC3 and stemness (CD44+/CD24-/low) indicate poor prognosis in triple-negative breast cancer. Hum Pathol. 2016;48:48–55.
64. Ray A, Vasudevan S, Sengupta S. 6-Shogaol inhibits breast cancer cells and stem cell-like spheroids by modulation of notch signaling pathway and induction of autophagic cell death. PLoS One. 2015;10(9):e0137614.
65. Galavotti S, Bartesaghi S, Faccenda D, Shaked-Rabi M, Sanzone S, McEvoy A, Dinsdale D, Condorelli F, Brandner S, Campanella M, Grose R, Jones C, Salomoni P. The autophagy-associated factors DRAM1 and p62 regulate cell migration and invasion in glioblastoma stem cells. Oncogene. 2013;32(6):699–712.
66. Angeletti F, Fossati G, Pattarozzi A, Würth R, Solari A, Daga A, Masiello I, Barbieri F, Florio T, Comincini S. Inhibition of the autophagy pathway synergistically potentiates the cytotoxic activity of givinostat (ITF2357) on human glioblastoma cancer stem cells. Front Mol Neurosci. 2016;9:107.
67. Yang S, Wang X, Contino G, Liesa M, Sahin E, Ying H, Bause A, Li Y, Stommel JM, Dell'antonio G, Mautner J, Tonon G, Haigis M, Shirihai OS, Doglioni C, Bardeesy N, Kimmelman AC. Pancreatic cancers require autophagy for tumor growth. Genes Dev. 2011;25(7):717–29.
68. Endo S, Nakata K, Sagara A, Koikawa K, Ando Y, Kibe S, Takesue S, Nakayama H, Abe T, Okumura T, Moriyama T, Miyasaka Y, Ohuchida K, Ohtsuka T, Mizumoto K, Nakamura M. Autophagy inhibition enhances antiproliferative effect of salinomycin in pancreatic cancer cells. Pancreatology. 2017;17(6):990–6.
69. Yang MC, Wang HC, Hou YC, Tung HL, Chiu TJ, Shan YS. Blockade of autophagy reduces pancreatic cancer stem cell activity and potentiates the tumoricidal effect of gemcitabine. Mol Cancer. 2015;14:179.
70. Peng Q, Qin J, Zhang Y, Cheng X, Wang X, Lu W, Xie X, Zhang S. Autophagy maintains the stemness of ovarian cancer stem cells by FOXA2. J Exp Clin Cancer Res. 2017;36(1):171.
71. Yang Y, Yu L, Li J, Yuan YH, Wang XL, Yan SR, Li DS, Ding Y. Autophagy regulates the stemness of cervical cancer stem cells. Biologics. 2017;11:71–9.
72. Ran X, Zhou P, Zhang K. Autophagy plays an important role in stemness mediation and the novel dual function of EIG121 in both autophagy and stemness regulation of endometrial carcinoma JEC cells. Int J Oncol. 2017;51(2):644–56.
73. Ojha R, Jha V, Singh SK, Bhattacharyya S. Autophagy inhibition suppresses the tumorigenic potential of cancer stem cell enriched side population in bladder cancer. Biochim Biophys Acta. 2014;1842(11):2073–86.
74. Ojha R, Jha V, Singh SK. Gemcitabine and mitomycin induced autophagy regulates cancer stem cell pool in urothelial carcinoma cells. Biochim Biophys Acta. 2016;1863(2):347–59.
75. Li J, Hu SB, Wang LY, Zhang X, Zhou X, Yang B, Li JH, Xiong J, Liu N, Li Y, Wu YZ, Zheng QC. Autophagy-dependent generation of Axin2+ cancer stem-like cells promotes hepatocarcinogenesis in liver cirrhosis. Oncogene. 2017;36(48):6725–37.
76. Li J, Chen JN, Zeng TT, He F, Chen SP, Ma S, Bi J, Zhu XF, Guan XY. CD133+ liver cancer stem cells resist interferon-gamma-induced autophagy. BMC Cancer. 2016;16:15.
77. Wiener Z, Högström J, Hyvönen V, Band AM, Kallio P, Holopainen T, Dufva O, Haglund C, Kruuna O, Oliver G, Ben-Neriah Y, Alitalo K. Prox1 promotes expansion of the colorectal cancer stem cell population to fuel tumor growth and ischemia resistance. Cell Rep. 2014;8(6):1943–56.
78. Zhai H, Fesler A, Ba Y, Wu S, Ju J. Inhibition of colorectal cancer stem cell survival and invasive potential by hsa-miR-140-5p mediated suppression of Smad2 and autophagy. Oncotarget. 2015;6(23):19735–46.
79. Singla M, Bhattacharyya S. Autophagy as a potential therapeutic target during epithelial to mesenchymal transition in renal cell carcinoma: an in vitro study. Biomed Pharmacother. 2017;94:332–40.

80. Fang D, Kitamura H. Cancer stem cells and epithelial-mesenchymal transition in urothelial carcinoma: possible pathways and potential therapeutic approaches. Int J Urol. 2018;25(1):7–17.
81. Yang Y, Fan Y, Qi Y, Liu D, Wu K, Wen F, Zhao S. Side population cells separated from A549 lung cancer cell line possess cancer stem cell-like properties and inhibition of autophagy potentiates the cytotoxic effect of cisplatin. Oncol Rep. 2015;34(2):929–35.
82. Li X, Wu XQ, Deng R, Li DD, Tang J, Chen WD, Chen JH, Ji J, Jiao L, Jiang S, Yang F, Feng GK, Senthilkumar R, Yue F, Zhang HL, Wu RY, Yu Y, Xu XL, Mai J, Li ZL, Peng XD, Huang Y, Huang X, Ma NF, Tao Q, Zeng YX, Zhu XF. CaMKII-mediated Beclin 1 phosphorylation regulates autophagy that promotes degradation of Id and neuroblastoma cell differentiation. Nat Commun. 2017;8(1):1159.
83. Galluzzi L, Bravo-San Pedro JM, Levine B, Green DR, Kroemer G. Pharmacological modulation of autophagy: therapeutic potential and persisting obstacles. Nat Rev Drug Discov. 2017;16(7):487–511.
84. Denton D, Xu T, Kumar S. Autophagy as a pro-death pathway. Immunol Cell Biol. 2015; 93(1):35–42.
85. Fulda S, Kögel D. Cell death by autophagy: emerging molecular mechanisms and implications for cancer therapy. Oncogene. 2015;34(40):5105–13.

The manufacturer's authorised representative in the EU is Springer Nature Customer Service Centre GmbH, Europaplatz 3, 69115 Heidelberg, Germany. If you have any concerns regarding our products, please contact ProductSafety@springernature.com

Printed and bound by CPI Group (UK) Ltd, Croydon, CR0 4YY
30/09/2026
02225238-0001